Take off Your Hoop Earrings Before Putting on Your Gas Mask

Take off Your Hoop Earrings Before Putting on Your Gas Mask

A Civilian Chick's Guide to Surviving a War Zone

Traci Scott

©2024 All Rights Reserved. No portion of this book may be reproduced, stored in a retrieval system, or transmitted in any form or by any means-electronic, mechanical, photocopy, recording, scanning, or other-except for brief quotations in critical reviews or articles without the prior permission of the author.

Published by Game Changer Publishing

Paperback ISBN: 978-1-963793-41-3
Hardcover ISBN: 978-1-963793-42-0
Digital: ISBN: 978-1-963793-43-7

www.GameChangerPublishing.com

DEDICATION

Thousands died in the War in Iraq. I dedicate this book to those who sacrificed their lives to help rebuild the country. Of those, these kind and dedicated heroes personally touched my life: Colonel Chad Buehring, François de Beer, Command Sergeant Major Jerry Wilson, Fern Holland, Salwa Oumashi, and Bob Zangas.

I offer my deepest love to you, Daddy, Mom, Lala, and PopPop.

Paula, Stephanie, Susan, Kristi, and Fahmi, you have left a profound impact on me, and for that, I love you very much.

Read This First

Thank you for buying and reading my book, I would like to keep the conversation going!

Scan the QR code to get connected

Take off Your Hoop Earrings Before Putting on Your Gas Mask

A Civilian Chick's Guide to Surviving a War Zone

Traci Scott

www.GameChangerPublishing.com

"You have enemies? Good!
That means you stood up for something, sometime in your life."
– Victor Hugo

Table of Contents

Preface 1

Chapter 1 – When You Go to a War Zone, Be Ready for Friendly Fire 5

Chapter 2 – Take off Your Hoop Earrings Before Putting on Your Gas Mask.......... 21

Chapter 3 – Be Nice to the Mean Girl Who Greets You When You Arrive: You Never Know When You'll See Her Again... 27

Chapter 4 – War Is Hell: When You're Going Through Hell, Keep Going! 37

Chapter 5 – My Friend Fahmi... 51

Chapter 6 – BBFs (Bunker Buddies Forever) Are Like BFFs: You Gotta Have Good Girlfriends to Get Through a War... 59

Chapter 7 – Give an Iraqi Boy a Toy and Pray He Doesn't Trade It for a Rocket Launcher.. 65

Chapter 8 – It's Not Smart to Go Hitchhiking in the Middle East, Especially During a War... 71

Chapter 9 – Honey, Er, Sir, I Think I Blew Up Our Car 79

Chapter 10 – If Your Tour Guide Tells You Jesus Blessed the Water, Remember… That Was Two Thousand Years Ago.. 85

Chapter 11 – Sometimes, You Just Gotta Deal with Geraldo Rivera........................ 91

Chapter 12 – If the Soldier Tells You There's No Room on the Bus, Don't Sneak Through the Back Door... 95

Chapter 13 – Never Ever, Ever Cry at Work—Ever! 103

Chapter 14 – It's Not a Good Idea to Wear Flip-Flops to Work: You Never Know When Your Building Will Blow Up in Front of You 109

Chapter 15 – Always Be Prepared in a War Zone Because You Never Know When You're Gonna Have to Put Makeup on the Secretary of Defense 115

Chapter 16 – Epilogue... 121

Preface

*"I didn't make the decision to start a war in Iraq,
but when my country called, I chose to go."*

If you are reading this book expecting my opinion on whether or not the United States should have gone to war in Iraq, close it right now and put it back on the shelf (or hit the delete button on your Kindle). This book is not about that. Nor is it an intellectual analysis of the history of U.S.-Middle East relations. This book is about some of my experiences as a civilian supporting the U.S. military in Operation Iraqi Freedom, or OIF.

Furthermore, I've recounted a series of experiences about which I've had some time to reflect, and I've realized there are larger life lessons that I want to share with those who are going into their own combat zones, wherever they may be. I chose to go to war even though I didn't know how to shoot a gun and didn't have any experience rebuilding a country. While I didn't have those skills, I did have other things to offer. I had years of experience in the media and dealing with the press, and I thought, *If that's all I can offer, then I will offer it and do what I can.*

It's been nearly six years since I left the bombings, IED-laced roads, and bullets in Iraq, yet I sit here torn over whether I should continue... with writing this book, that is. I just don't want to be a target all over again.

So many books and movies have been written by soldiers who were shot, blown up, or had to kill others during OIF. Many have written unbelievably courageous stories of surviving physical and emotional wounds.

In some cases, family members of those killed tell their stories as a healing balm to deal with the painful loss. One of those is Darrell Griffin. When his son Darrell Jr. (Skip) went to Iraq, they exchanged emails and promised to turn what they wrote into a book. During Skip's second tour in 2007, he was shot and killed by a sniper. Darrell Sr. was determined to follow through on his mission. He made his way to Iraq (a feat in itself since civilians were not allowed to travel to the theater of operation at the time) to meet the soldiers who served with Skip when he was killed. The trip not only allowed him to finish the book but also to make peace with his son's death.

Books like Darrell Sr.'s, which share incredible journeys through pain and loss, are so compelling and necessary. And then here I am with my book, which has to do with things like wearing hoop earrings while trying to stuff a gas mask over my head... But, before you pass judgment, read through the book. You will see that I also endured painful moments, whether I was wearing hoop earrings or not.

Then there's Army Captain Joshua Mantz's story. He was shot in the leg in Iraq and was actually dead for 15 minutes before being revived. He went back to Iraq to finish his mission and, more importantly, face his fears. I don't want this book to dishonor Josh or the soldiers who died serving alongside him. As a matter of fact, I asked Josh to read this to make sure that he wouldn't be offended.

As I've said, this is a different book about the war in Iraq. More than that, it's a teaching tool, not for historians but for young girls and even young men who are new to the combat zone. I went to the 10,000-foot level, looked back, and said, "I want to share a few stories and the life lessons that come with them."

The foremost lesson is that life is a process. Whether your battlefield is in the Middle East, in the middle of a boardroom, or facing a cancer diagnosis, you have to go through the process one step at a time. Furthermore, please don't make the same mistakes I did. You will save yourself a lot of pain. And one final lesson: in life, there are times when you must have faith beyond your understanding and courage beyond your expectations to get you through them.

Sometimes, the battle in your mind can be just as difficult to navigate as a minefield. What do I mean by that? You know the voices in your head. They tell you, *Go this way. No, go that way. If I go down this road, where's the IED?* Or, *You idiot, if you had taken that other road, you wouldn't have gotten into this mess.*

So, getting back to my original point, I struggled with writing this book because I wasn't on the frontlines shooting an M-4 or standing watch over an FOB (forward operating base). I didn't know how to shoot a gun or rebuild a country, but I had my own experiences and wanted to use my talents to do my part to help.

Why did I go? Because I wanted to serve my country. My daddy, Jesse H. Scott, Jr., a Tuskegee Airman, and my stepdad, "PopPop," also an Air Force veteran, implored me not to join the military, but they didn't say anything about going to the war as a civilian. Also, I had wanderlust. I was fascinated by the Middle East and intrigued by its people, culture, and religions. But most of all, I wanted to do my part to help. Sometimes, I justify it by saying, "If you pull up to a car wreck and somebody's lying there hurt, you don't just run over to them and start yelling at them about who caused the accident. You ask them where they are hurt and try to comfort them. You leave it up to the investigators and crash experts to figure out who's at fault."

Before I went to Iraq, I was so naïve. One dear friend used to tell me I was Pollyannaish and that once I hit Iraq, I wasn't going to be in Kansas anymore (funny how I ended up in Kansas anyway). By the time I left Iraq after nine months in 2004, I was hard, aloof, bitter, angry, impatient, dark, and confused, definitely not the characteristics used to describe sweet little Pollyanna.

You've heard the saying, "What doesn't kill you makes you stronger." Iraq nearly killed me. You can only go so long enduring friends being killed, dodging mortars, and witnessing endless carnage before things take their toll. Different people come through a battle with different scars. The ones on the body are easy to see, but the scars on the psyche and the soul can remain hidden and surface unpredictably: when a flag billows in the breeze, when someone cuts you off in traffic, or even during a thunderstorm.

Wars can change anyone, including an idealistic young woman who decides she simply wants to throw on her hoop earrings and follow the other soldiers into battle.

CHAPTER 1

When You Go to a War Zone, Be Ready for Friendly Fire

No one prepared me for the lesson I learned when interacting with battle-hardened warriors. Here's the skinny: some men don't believe in sending women into a combat zone. When I ran into them, it was like getting nipped in the rear end by friendly fire. So you better toughen up and bring your big girl panties, or you will be dismissed by them!

When I was a little girl, and someone asked me what I wanted to do when I grew up, going to a war zone was not on the list. But there I was, standing in a Capitol Hill corridor on my cellphone, rather stunned that someone had said that our country needed my help. It was 2003.

First of all, I am extremely naïve, not exactly soldier material, even though I was an Air Force brat. Despite this, I knew it would be dangerous, and I could lose my life. But as I stood there on the phone, staring out into the courtyard of the Longworth Building, my mind was racing a zillion miles an hour with the thoughts one thinks before making a life-altering decision. In my heart, though, I already knew what I was going to do. I had to go.

I knew it wouldn't be easy getting into a war zone. After all, Iraq was more than 10,000 miles and six time zones away. Little did I know that my biggest hurdle would pop up before I could even get my TDY (temporary duty) orders.

I will start back at the beginning. Why on earth did I even want to go to a war zone? As hokey or fake as it may sound, I really wanted to serve my country. I worked for someone a few years back who impressed upon me the importance of service to the country you love. If you love this country, you must do your part to help, no matter how large or small that part is. I was also up for an adventure; I like to be where the action is, an impulse from my TV days. I was absolutely fascinated by the Middle East and Islamic culture. I didn't understand the Muslim world, and my curiosity led me to want a greater understanding. That's why I went to Iraq—nothing more.

While I was overseas, rumors and press stories claimed that most of the staffers in the Strategic Communications office (Strat Comm) in the Green Room were there to continue the "Bush agenda" and curry favor with the administration to get a job when they returned home. That was not me. Let me repeat that—I did not go to Iraq to get a position in the Bush administration. As a matter of fact, I lost my job on the Hill when I went to Iraq. I didn't have a job when I came home and was unemployed for about three months. I did end up getting a political appointee slot two years later, but I want you to understand that was not my plan.

As I've said, I am a very naïve person. I want to believe that everyone is good and has the best intentions for all. I learned in Iraq that this is not necessarily true. I know it's a shame that I was nearly forty years old before I transformed from a "Pollyanna" to a "Baghdad Bitch."

In March 2003, at the time of the infamous "Shock and Awe" bombing campaign—the opening firestorm that started the war in Iraq—I was working as the communications director for Nevada Congressman Jon Porter. During his first few months in office, he was invited to an event at the Pentagon for new members of office. While there, I met a young, energetic press operative named Claude Chafin. Claude's job was to act as a liaison between the secretary of defense's Legislative Affairs Office and press secretaries in the House and Senate. It was a tough job that needed someone with Claude's patience and unique ability to let things roll off his back.

After our first meeting at the Pentagon, Claude became a source of information when it came to Defense issues, but more than that, we became friends. In the days after the beginning of the war, Claude and I stayed in contact, and he told me that he was going to make a trip to Iraq for a couple of weeks to help the newly formed ORHA (Organization of Reconstruction and Humanitarian Assistance), the predecessor to the CPA, in their press operations center. I wished him the best of luck but secretly envied his ability to just pick up and head off to the Middle East.

One morning, Claude phoned me from a place he called "the palace" to tell me how things were going. It was unbearably hot, so hot that the people there had chosen to sleep outside since the palace had lost its AC during the initial bombing. I was more intrigued by his stories than by what I had seen on TV. Then he dropped a bombshell of his own: they needed press officers over there.

Claude was good at selling me on the idea. I was partly sold on it already, but when he said, "Your country needs you," his words hit me like a shot of adrenaline. I was ready to go as soon as I hung up the phone. I told him we would talk about it more when he returned home.

True to his word, when Claude returned to the Pentagon, he urged me to come on board and join in the fight—from the civilian side. I was hesitant to pull the trigger. First, I had to get past my mom, Shirley Jones. My mom was my best friend. We were so tight; I used to say I never really left the womb. We talked every day, and I lived about 20 minutes from her house. I went home every Sunday for dinner, called her for advice when I needed to decide something, and was jealous of anyone who monopolized her time. When I showed up at the house, if she was on the phone, I wanted her to hurry and get off so I could have my "mom" time. Yes, it was all about me—and my mom spoiled me to death! In my heart, I wanted to go to Iraq, but how on earth was I going to cut my umbilical cord without causing severe damage?

Then the thoughts started swirling in my head... *This is your chance to serve your country... You can give back... You can finally travel to the Middle East... If you don't go, then who will? Why not you?* When my mind finally settled, I had made my decision. I was going to go to Iraq. I called Claude.

I finally called Claude and told him that I had made up my mind. I was going to Iraq. He was gleeful. I, on the other hand, dreaded dropping the bomb on Mrs. Shirley Jones. I had to come up with a strategy. I called my dear friend Paul Taylor (P.T.), with whom I had worked for years at CBS News. P.T. directed the Sunday show *Face the Nation*, and he was also dear friends with my mom and PopPop, my stepdad Joe Jones. P.T. knew my mom well enough for us to figure out how to break the news to her gently.

One afternoon, I sat down in the family room with my PopPop and Mom and just blurted it out: "Mom, I am going to Iraq." As expected, her reaction was a lot like the "Shock and Awe" bombing campaign. She exploded!

"You're what?" she exclaimed. "No, you are not!" Then she went into a tirade. It was ugly. But my mind was made up, and just like her, when I made up my mind, there was no turning back.

Mom and I didn't speak to one another for three days. For us, that was an eternity since we generally spoke every day, sometimes two or three times a day. But I had to understand that she could potentially lose her "baby" and only daughter. It was pretty extreme.

One day, she came around. She called me and matter-of-factly asked me if I had all the shots I'd need to go overseas. I was so relieved. While she couldn't come out and say the words, "I support you and your decision," she did the best she could at the time. From that day on, anything I did on my way to or while in Iraq, she did her best to help me with—from making sure she gave me her best advice to the care packages she sent filled with goodies and her love.

So, after going through a little self-doubt, a series of immunizations, and Shirley Jones, it was time to head off to Iraq. I thought it would be easy to get on a plane and go, but Claude said that before I left, I had to meet with someone in the Pentagon who would approve my slot. At the time, I didn't understand what the White House Liaison Office was or why I had to go, but that was part of the process. I thought, *Well, I'll just go and get my orders and then get out of Dodge.*

At the office, I dropped off my resume and spoke to a gentleman about my altruistic ambitions for going to a combat zone. As I was leaving, he lobbed the first volley of "friendly fire" that I would experience: "You know, I don't believe in sending women into war." I stood there, floored. Was he joking? I didn't see a hint of a chuckle anywhere. I didn't know what to say, so I didn't say anything. I just smiled and left. But the words were seared into my brain.

Here's a quick history lesson. Women had been supporting the U.S. Army since the Revolutionary War in 1775, doing the jobs soldiers didn't have time to do, like nursing, washing uniforms, and cooking (www.army.mil/women). It wasn't until 1942, during World War II, that women could serve in the Army, when Congresswoman Edith Nourse Rogers of Massachusetts introduced a bill to create a women's corps.

This new women's corps was different than the nurses' corps that had already been established. The Women's Army Auxiliary Corps came first, and eventually, it developed into the Women's Army Corps, or WACs, as they were called. In the Pentagon basement, there is a display honoring the first women in the WAC program. The first class consisted of 440 women, including 40 Black women, who, in line with the times, were segregated into their own separate unit.

Fast forward to Operation Iraqi Freedom, and female soldiers were in the battle. The Iraqi battlefront was one of the first theaters of operations where U.S. female soldiers carried the weight of war alongside the men. They were in the fight: shooting, getting shot at, and losing their limbs and lives. Many left their babies, children, and husbands behind, all in the name of freedom. As of April 2008, 99 female soldiers and 13 U.S. civilian women were killed in Iraq. I honor those women and their sacrifices.

Back to my civilian experience. After that first "shot across the bow" from the Liaison's Office, I knew I would be around a bunch of men in Iraq who didn't want me there. I accepted this, but that didn't mean I had to wither away like a wilting wallflower. Some people believed I had the experience and talent to get the job done, and I was determined to do my best.

Just a few weeks later, I would run into more friendly fire. My job in Iraq was to serve as the "special projects coordinator" with the Coalition Provisional Authority (CPA), which meant I worked with VIP journalists who wanted to do stories on the rebuilding of Iraq.

When journalists reported on military operations and battles on the frontlines, their stories were filtered through a military information center. In Iraq, this center was known as the CPIC (pronounced "see-pick"), an acronym for the Combined Press Information Center. Some journalists were also part of the embedding program. In this program, journalists would hook up with a military unit and stay with them day and night for a couple of days or weeks at a time. It was an excellent way for reporters to live, eat, and breathe the war.

However, if a reporter wanted to do a story on how the Coalition Provisional Authority was working with the Iraqis to rebuild their government and services, they came through our office, Strat Comm. Now, there has been a lot of criticism of the CPA, and I am not going to get into all of that in this book. The purpose of this book is to tell you about some of the junk I went through in Iraq in hopes that you will use the lessons in your own war zone.

In November 2003, my second big assignment was to escort MSNBC Correspondent Bob Arnot (aka Dr. Bob), producer Noah Oppenheim, and their crew on a trip up north. I knew of Dr. Bob when I worked at CBS. He was a network correspondent who focused mostly on health issues for the network and was highly intelligent. He spoke about 11 languages, was a pilot, and was a pretty nice guy. Okay, back to my story. Dr. Bob wanted to go to Kirkuk and Mosul for a story on how the soldiers were working with Iraqis to rebuild prisons and create good relations with the locals. He also wanted to go to Mosul to meet Army General David Petraeus, a major general at that time, which means he had two stars, and to talk with him about the successes the 101st Airborne was having in securing that part of the country.

To get Dr. Bob, Noah, and crew around the country, I had to coordinate with the military folks in the palace. They would make sure we had air support (Blackhawk helicopters) to fly us up to Kirkuk and then on to Mosul. I had only been in Iraq for a few weeks and was looking forward to getting out of the office, out of the Green Zone, to see Iraq.

Now, I had never been on a military aircraft. I had been on a civilian chopper as a reporter in Vegas and enjoyed being able to look down on the scenery below. But flying in a military chopper is a completely different experience. First of all, Blackhawk helicopters are very loud. Second, you are usually flying with the doors open. Third, a gunner is posted in the open window, ready to shoot at any potential threat. Not exactly what I was used to in Vegas.

We left Baghdad on our way up to Kirkuk, and I admit I was pretty nervous. I was about to take a high-profile correspondent and his crew on my first trip out of the Green Zone up to northern Iraq. I thought, *How did I get here? How did I get to the point where I ordered a chopper, got bulletproof vests, and am now on my way over the desert to northern Iraq?* But I had to stuff all the fear and wonderment deep down inside. I couldn't show that I was afraid or display a lack of confidence in any way at all. I had to forge ahead and do my job.

The Blackhawks picked us up at the landing zone (LZ) in the Green Zone and flew us up to Kirkuk. As I said, this was my first airlift out of Baghdad, and I was absolutely fascinated by what I saw. First, Baghdad looks like L.A. on a bad smog day. More than six million people lived in this city of stone homes, along with some cement houses and mud huts. What completely floored me, though, was that almost every roof we flew over had a satellite dish. When Saddam Hussein was in power, he banned satellite dishes. Once he was removed, people snatched up satellite dishes like they were free. The saddest-looking shack with sheep running around the yard had a satellite dish on the roof—go figure.

Outside the city, we encountered a pleasant surprise: groves and groves of palm trees. I love palm trees, but I didn't expect to see so many of them. Our chopper glided over them so low that I could almost jump down and bounce on them like a green trampoline.

Baghdad was a dirty city without much vegetation, but out in the countryside, we could've been flying over Florida. As we flew further north, it was like we were entering another country. There were beautiful trees, acres of green grass, and sloping hills—a green oasis. We flew over a crystal blue lake in the middle of a green valley, and I thought we were in paradise. I had never known Iraq could be so beautiful!

We finally touched down at our first stop in Kirkuk. There, we would meet with a group of U.S. contractors who were helping the Kurdish people rebuild their firefighting capabilities to deal with the burning oil fields. Kirkuk is full of oil fields. The Kurds there in the north were fighting the Iraqis in the southern part of the country over who would reap the financial benefits of the oil the fields produced. That's the short story. If you want to learn more about Kirkuk, Google it.

We spent the afternoon visiting the oil fields, talking with some of the folks who ran the oil companies, and interviewing the contractors with the fire equipment. Before we knew it, it was time to go to our second location in Mosul.

When you are traveling in Iraq via military transportation, there can be a bit of a culture shock. First, you are dropped off at the LZ. Then, instead of the chopper hanging out and waiting for you, the pilot might take off and fly to another location. Or, while you are waiting for your ride to your next location, sometimes it will show up at the appointed time, and sometimes it won't. No matter what, though, you better have your butt there when the choppers show up because all they do is touch down, pick up their load, and take off. There's no waiting around. If you miss them, you are stuck for who knows how long.

This happens day or night. If it's night, you're standing at the LZ in the dark, hoping nobody starts shooting. It's kind of cool in the dark; you can hear the choppers, but because the pilots keep their lights off and use night-vision goggles, you can't really see them until the wind from their blades all but knocks you over.

Our next stop was Mosul. Dr. Bob wanted to go on patrol with soldiers through a market, interview troops, and then try to get an interview with MG Petraeus. When we arrived, we were starving. We'd been flying all day, walked around oil fields, and had eaten very little. Unfortunately, we had arrived at the 101st Airborne headquarters in one of Saddam's palaces after the chow hall had closed.

We made peace with the fact that we weren't going to eat and were on our way to a briefing area for an update on Mosul when a top soldier, then Command Sergeant Major Jerry Wilson, came around the corner from the chow hall—with an armful of food! We were so happy! Now, hang with me 'cause I am telling this story for a reason. When you have soldiers like that take care of you, you really appreciate it. And just like with everything in Iraq, you never know when you are going to see somebody again, so when they do take care of you, you offer your deep appreciation and remember to return the favor.

About two weeks after we visited Mosul, CSM Wilson was brutally killed by insurgents. They grabbed him and Specialist Rel A. Ravago out of their Humvees, slashed their throats, and dragged their bodies through the streets of Mosul.

I thought a lot about the sergeant major and what his family had to go through when they found out the details of his vicious murder. This battlefield was bloody and unfair. Insurgents and other bad guys would kill our soldiers, drag their bodies through the street like it was Mardi Gras, and get away with it. Meanwhile, our soldiers were put on trial for mistreating Iraqi citizens. Don't get me wrong: wrong is wrong is wrong. But if our soldiers had grabbed two Iraqis, slashed them up, videotaped it, and paraded their headless bodies through the streets, they would have been tried for war crimes and sent off to prison at the very least. I know I am on my soapbox, but war is bloody, vicious, and unfair.

Back to my story.

We went in to be briefed on the fight in Mosul by one of the commanders in the JOC (Joint Operations Center). When we walked in, it was like we'd entered a cross between a high-tech movie studio and a World War II command center. There were TVs all around, computer monitors with cryptic information, and rows of wooden desks behind the main control area. We were introduced to an officer with a slight but sturdy build. He was a little guy, but his crew cut and quick reflexes said he would cut you in one minute and then sit down for a cup of coffee in the next. He was tough.

Our military escort told us General Getter would be conducting our "ops briefing." We were directed to a row of seats behind him and settled ourselves in them for his slide show. As he was going through his brief, I noticed he would glance at me with a look just short of a scowl. I thought, *Here we go, an officer who doesn't especially appreciate my presence.*

Have you ever had that happen to you? You walk into a room, and instead of everyone smiling and putting out their hands to shake, they look down on you like they are thinking, *What the "f" are you doing here?* Well, that's how the situation felt to me. I became self-conscious and tried to focus on the briefing slides. But I kept thinking, *Why is he grittin' on me?* Dr. Bob brought me back to earth when he asked me to help him with a piece of equipment. This allowed me to focus on a task, so I didn't worry about the eyes boring into me.

Today, I can still see Gen. Getter swiveling around in his chair to stare at me. I kept thinking, *Is it because I am with the CPA, or is it the fact that I'm a civilian chick, or maybe it's the hoop earrings?* Whatever it was, I was going to have to stick to the task at hand and focus on my work.

After our briefing, Dr. Bob and his crew wanted to set up for a live shot. Yes, it was 10:30 at night. We'd been up since the crack of dawn, and there wasn't any chance we were going to go to bed anytime soon. That's just the way it is in a war zone.

That night, we spent what few hours of sleep we had in some crusty-busty former hotel. We turned on the faucet, and something close to mud came seeping out. It was obvious we weren't going to have a concierge fix that one. And so, we bunked down for the night in a couple of cots and tried to sleep for a few minutes. However, I was so keyed up and so freaking nervous that I couldn't sleep at all. Our "hotel" room faced a busy highway, and all I could think about was some insurgent lobbing a mortar at the room.

After what felt like just a few moments, we packed up to head out for a morning patrol. We met up with Commander Dirksen and his unit to spend the day with them and learn more about their attempt to "win the hearts and minds of the Iraqis." Commander Dirksen took us to the Mosul Police Academy, where they were training Iraqis to be local law enforcement officers. We were just down the road from where U.S. soldiers had cornered and annihilated Saddam's evil sons, Uday and Qusay, a few months earlier.

Well, as I was standing there, listening to the briefing, I noticed Commander Dirksen giving me that hairy eyeball look. I thought, *What is this?* I was minding my own business, trying to do my work, and here was another officer looking at me like I didn't belong. I looked back at him and mentally shook my head. I was very aware that these soldiers had an extremely important job and didn't have time for civilian chicks on a battlefield, but I had my job to do as well, and that was making sure these journalists had the resources they needed to tell their stories. I stood my ground with Commander Dirksen, but I stayed out of his way.

Later that day, as we were trekking around Mosul, we headed to a prison that the soldiers had helped refurbish and were working with the Iraqis to "manage." Commander Dirksen received some devastating news. He walked over to us and told us he had just received a call that there was a *fatwa* on him. A *fatwa* is an Islamic decree, but we Westerners would call it a "hit." Commander Dirksen looked shaken. In the States, it's one thing to have somebody threaten to kick your ass, but in Iraq, it's a whole different level. Insurgents were notorious for following through on their threats, and we took them seriously. If they said they were going to get you, they did. So, we understood the gravity of the situation.

By the time we left Mosul, I wasn't sure what kind of impression I had left on the officers, but I knew I wouldn't forget them. Honestly, I hoped I wouldn't have to run into them again. Well, that didn't work for long. A few months later, I was back in the north on another trip.

By February 2004, I was ready for a break. I had been in Iraq for five months, lost a friend, nearly lost another, and gone through countless attacks on the Green Zone, and I needed to take some time off. Deputy Secretary of Defense Paul Wolfowitz had finished his third trip to Iraq and would soon be on his way back to the U.S. I heard there might be enough space on his C-17 for me to hitch a ride home. The rub was that I had to ride on choppers from Baghdad up to Mosul with the travel party to get on the plane flying back to the States. In Mosul, I would come face-to-face with the officers again. Oh, well. I was traveling with the DEPSECDEF, as we said in Pentagon lingo, so how bad could it be?

When our choppers landed in Mosul, we had to go to an event commemorating the rebuilding of a neighborhood and were sent to a police training facility. Well, somewhere along the way, we met up with Commander Dirksen. Though our first meeting had been less than cordial, I was glad to see he was alive in spite of the *fatwa* on him. To my surprise, his greeting was much better this time. Instead of the hairy eyeball, I received a very different reception, and he seemed happy to see me. We caught up on how things were going in northern Iraq and acted like we were old buddies. I was so relieved.

Toward the end of my tour in Iraq, I had another chance meeting with General Getter. He was down in Baghdad, visiting with another top general, and as they were walking out of one of the Iraqi buildings, I looked up and saw him. The next thing I knew, I heard, "Hey, Traci!" and he waved hello to me. I couldn't believe it. He came over, and we embraced, happy to find that we were both still alive. As with Commander Dirksen, it was a 180-degree change from our first meeting. I was surprised. General Getter told me he had changed posts, and he seemed genuinely happy. I was happy for him.

Some months later, after I left Iraq, I saw General Getter walking through the halls of the Pentagon. We were happy to see we both were alive and had made it out of the sandbox safely. One day, when we greeted each other in the hall, I took advantage of the goodwill. "Sir, when we were in Iraq, you were pretty hard on me. Why?"

I think my forthrightness caught him off guard. He grinned and said, "I wanted to see if you could handle it." That was all I needed to know. I wasn't a soldier, and he couldn't use my rank to assess my skills, so he had to put me through his own personal boot camp to see if I was true blue or just hanging out in Iraq to be cute. Once I passed his test, he realized that I knew my job and was serious about it, and the walls came down.

As the months went by, General Getter and I would often see each other in the hall. At one point, he mentioned that he had a family member who had taken ill, so I told him about my mom and her battle with cancer. Every time I saw him, he would ask, "How's your mom?" That always touched me. A good officer gains loyalty from his soldiers. General Getter was the same with me, and I will always deeply appreciate that.

The last general to put me through the wringer ended up saving my life. General "Stone" was tough, and he was smart. When he walked into the room, folks would feel like hiding behind their cubicle walls. His features were so distinct that if I described him, the whole world would know who I was talking about.

The general and I used to get into heated discussions about lots of things. One day, I finally said, "Sir, I don't care how many stars you have on your lapel. I am a human being, and I expect to be treated as such." I think he was taken by surprise by my chutzpah.

There's a reason I am not in the military, not solely because my dad and PopPop asked me not to join. It's because I hate people yelling at me and using their authority over me. I hate it. I like having my right to fight back. An enlisted soldier cannot get into a verbal sparring match with an officer and disrespect his or her authority. Doing so is an Article 15 (allows commanders to administratively discipline troops without a formal court-martial) and could get a soldier kicked out of the military. With my outspokenness and independence, I would probably get an Article 15 every other day.

Apart from verbal repartees with General Stone, things were not going well for me in Baghdad. My mom was sick, I had lost a couple of friends, I had been sick, and life, in general, just really sucked. One day, after General Stone and I had made peace in our own battle, he mentioned something about running. He went so far as to say that I didn't have the discipline to get up at 5:00 AM to go running with him. Not one to back down from a challenge, I took him up on his offer.

The next morning, I got up at five, threw on my running clothes, and went out in the dark to meet General Stone for a run. It was not what I expected. At about 5:20, I was standing at the back side of the palace, waiting for him, and at 5:25, he came running up the path with one of his military aides. When he got to where I was standing, I cheerfully greeted him, "Good morning, General Stone!"

"Hmmff!" was all I heard in reply, and that's how things went four days a week during my "run" with General Stone. I would stand there, he'd come running up, I'd say good morning, and he'd just run by me like I was a tree limb. I'd catch up with him for about 50 yards, and then he'd take off, leaving me in the Iraqi dust. But I loved it. I looked forward to taunting him in the morning with my cheery disposition. He hated it; General Stone was not a morning person.

After a few days, I noticed my attitude had started to change. I actually had something to look forward to in this crazy sandbox other than what time the bombing would start. Sometimes, while I was on my run, the Green Zone would get shelled, and I had to think quickly: *Do I keep running in the direction of the explosion or turn back around?* Most of the time, I ran in the direction of the hit. The insurgents had bad aim, and I figured they couldn't hit the same spot twice, so I continued on my way.

I was finally coming out of my dark place in Iraq, near to getting back on track. General Stone, unbeknownst to him, had saved my life by making me get up in the morning and forcing me to exercise. I had been right at the edge of a depression so deep that I don't know if I could have come out of it.

I am proud to say that General Stone and I are still very close friends today. We've gone full circle. I've met his wonderful family and keep in touch with him on a regular basis.

So, the lesson here, folks, is that in a war, you've got to prepare yourself for battle: keep your wits about you, arm yourself, go in with your head in the game, size up the enemy, and be ready in case there's any friendly fire.

CHAPTER 2

Take off Your Hoop Earrings Before Putting on Your Gas Mask

> *The Namesake for this book has a very simple message. Life is a process. You have to go through the process if you want to succeed, reach your goal, or survive a war zone. For example, if you want to get through high school, you gotta show up for class. That's part of the process...*

By the time I made it to Kuwait, I was beyond exhausted. I boarded the plane at about midnight at Dulles Airport in Virginia, flown about six hours to London in the morning, had a nine-hour layover, boarded a flight for another six hours, and landed in Kuwait sometime in the afternoon. I don't remember what day. I had been up for nearly two days straight, with no shower, no real meal, and just a few hours of bobblehead sleeping on the plane. At that time, I didn't know the benefits of Ambien and just had to suffer from sleep deprivation.

Then came one of my more interesting moments in the Arab world: bathroom culture shock. When I went into the women's bathroom at the airport, I opened the door to the stall, and all I could do was stand there and stare at the hole in the floor. I just couldn't figure out what the deal was with the little showerhead hanging on the side wall.

I stood there for another moment, and then, finally, my brain fired up a command: *Move!* I squatted, took care of my business, and reached over for the toilet paper. There wasn't any! *Damn*, I thought. *Now what am I supposed to do?* A few seconds later, I remembered I had some Handi Wipes in my bag. I know, it's gross, but sometimes, you gotta do what you gotta do.

When I walked to the sink to wash my hands, I looked over into another stall and saw the porcelain goddess—with a roll of toilet paper! How had I missed that—a Western-style toilet? It didn't take too many bathroom trips before I learned that Islamic women did their business and used the mini shower heads to clean themselves instead of toilet paper. Nevertheless, I always had a package of Handi Wipes with me wherever I went.

After the bathroom fiasco, I made my way over to the customs counter to officially enter the country. The Kuwaiti checkpoint guards were very intimidating. First, there is something enigmatic about their features: dark hair, dark eyes, thick, dark eyebrows, and a secretive look. Yes, I am a black woman and used to dark features, but their ruddy countenance and mystifying air were foreign and fascinating to me.

I handed the Kuwaiti customs officer my passport. He looked at it, looked at it again, and looked at it some more. He was taking a little too long for my liking. Next thing I knew, he called over another officer, who looked like an Arab sheik with a potbelly. They both started talking and looking at my passport. I felt a tingle in my armpits.

"Come with me," said the second officer, who turned out to be the Kuwaiti TSA supervisor. These were, of course, the first English words I had heard from them. I started to get scared.

The supervisor led me to a room where he told me to sit down. My brain finally shot into overdrive. I was being detained! I couldn't believe it! I was in the Middle East, more like the Middle of Nowhere, with these guys who could slice my head off without thinking about it, and no one would know I was even missing. I wanted to cry. I was sitting in a room by myself, with no idea of what was going on, no idea who to call, and no clue as to what to do. I could see the supervisor conferring with another officer, but it didn't matter. I couldn't hear them, and even if I could, I didn't know what the hell they were saying.

Then my cell phone rang. Thank God! It was one of my friends calling to see how I was doing. I turned into a blubbering idiot as I told him I was being detained in Kuwait and didn't know what to do. Even though the call was from thousands of miles away, I felt comforted by a caring voice. Just about that time, the TSA supervisor walked over and said, "You are okay to go." I nearly jumped up and gave him a hug! I learned later that I should have shown the Kuwaiti customs agents my military ID, not my passport. I could have used that little nugget of news before leaving D.C.

During that time of the war, Americans going to Iraq were all housed at the Kuwaiti Hilton. It's all that you would imagine: a beautiful resort right on the Persian Gulf. Soldiers returning from a short vacation (R&R) would go there for a few days to decompress before returning to the fight. For civilians, it was where we in-processed and made last-minute preparations before heading off to war.

When I arrived at the Kuwaiti Hilton, I was toast. I went to the check-in room, got my room key, and was instructed to start my in-processing. Start in-processing? I'd been traveling for three days, and I was stanky dirty. My brain was so fried I couldn't spell my name, and they wanted me to start in-processing? I was dying for even a few minutes of sleep.

I dropped my stuff off in the room and then went to the restaurant and finally had a decent meal. I was relieved to see other Americans there but was too tired for a lucid conversation. They did tell me I didn't have to do the in-processing until the next day, so I went back to my room and passed out.

The next day was my first full day in Kuwait. I spent the entire day taking classes and learning the processes and procedures of the infamous Green Zone. One of the first things we had to do was get fitted for our gear. Yes, even though I was a civilian, I was issued fatigues, boots, and a helmet—just about everything soldiers had except for a weapon.

I went to the "dressing room" with some other civilian chick, who was not exactly what I expected to see heading into a war zone. Her name was Paula, and she was a bubbly, blonde, gregarious young woman, full of life and exuding good vibes and fun. Paula and I became best friends in the time it took us to find a uniform and boots that fit. As we rifled through the pants, we started talking. We talked and talked and talked. After getting our boots, we walked out together and stayed together all the way to Iraq.

That afternoon, Paula and I had some extra time, and she decided to go for a swim in the Gulf. Now, I was still pretty exhausted. I really wanted to go back to my room and try to catch up on the three days of sleep I'd missed. Paula, however, convinced me to go swimming instead. I think she said something like, "How many times can you say you went swimming in the Persian Gulf?"

The water was a beautiful shade of turquoise and was delightfully warm. Now, I can barely float. I waded in about waist-deep while Paula ventured a little further out into the crystal blue water. After a few minutes in the water, we decided we'd been out there long enough and headed back to shore, proud that we could mark "Swim in the Persian Gulf" off the bucket list. When we got to the hotel, someone who had seen us splashing around said, "You know there are floating mines out there, don't you?" As in the kind that explodes when something hits them. No wonder we were the only ones out there.

We decided to hang out at the pool instead. While Paula and another new friend, Jason, enjoyed soaking up the sun, I took the opportunity to find a lawn chair and take a nap. That was a mistake. I was just falling into a deep sleep when Paula woke me up and told me we had to go to another class—gas-mask training.

Gas-mask training was essential for every soldier and civilian going to war in Iraq. You may remember the early days of the war, pictures and video of warning sirens going off, and TV correspondents jumping for their gas masks. (Saddam Hussein earned his notorious reputation when he gassed his own people in the tiny northern village of Halabja. I eventually traveled there and saw pictures of the atrocities.) With this in mind and not knowing where Saddam was or what the insurgents were capable of pulling out of their sacks, we all had to be prepared for chemical warfare.

On our way to the classroom, a sudden wave of exhaustion hit me like a ton of bricks. I felt like I could've fallen asleep in mid-stride. My dear friend Paula had been in Kuwait a day or so before me and had overcome the jet lag. I, on the other hand, was barely functioning.

When we got to gas-mask training, I had to fight to remain awake, and that was not cool. This was the most important instruction we would receive before heading to Iraq. A few minutes into the class, it was apparent that I was not the only one still suffering from sleep deprivation.

We all went into a small conference room and sat, listening as the instructor explained how to put on our gas masks. Now, on a good day, I have average intelligence, but fighting exhaustion while trying to process the teacher's instructions about the intricate details of the proper way to pull the gas mask out of its package and put it on your face while under attack (all in a matter of seconds) was more than my little brain could handle.

I grabbed the back of the mask, stretched out the straps, and tried to pull it down over my head. "Ouch!" I nearly ripped my hoop earrings out of my earlobes! My friend Paula must have sensed that I was struggling, and at about the same time, she looked over and snapped, *"Traci! You gotta take off your hoop earrings BEFORE you put on your gas mask!"* At the time, it was a simple moment of absent-mindedness, but for me, it became a metaphor for life. Life is a process, and you can't get to where you want unless you go through the process.

At about that time, the instructor looked over the class and saw that all of us were pretty bleary-eyed. "Okay," he bellowed. "I can tell this is not sinking in. All of you go back to your rooms, get some sleep, and we will do this again later." I guess we all flunked gas-mask training.

Well, I did learn one lesson that day: leave the hoop earrings off… at least when you have to put an object over your head designed to squeeze over your ears and cut out all outside air. You can always put them back on once the device is properly in place.

By the way, I never did have to use the gas mask—thank God.

CHAPTER 3

Be Nice to the Mean Girl Who Greets You When You Arrive: You Never Know When You'll See Her Again

> *Always remember, first impressions aren't necessarily correct. And be nice to everyone because you never know when you might need their help.*

I had all these questions about what Iraq would be like when I finally made it there. Would it be nasty dirty? Would it stink? What were the people like? Who would I work with? What would the soldiers be like? Would I see anybody get shot? Most of all, I wondered if I would come back alive and whether I'd have all my limbs when I returned. I struggled with these questions daily before I left for Iraq, and while I was there, they never drifted too far from my mind.

The images of the war on TV gave me some idea of what things were like in Iraq. Mostly, it looked brown, dusty, and dirty. But TV spots can't express the smell of a place—or the heat. I also remember images of people looting stores and soldiers standing around with their weapons poised as if unsure of what to do. Along with my obsession with what the combat zone would be like, I had this weird fantasy that as soon as I stepped off the plane, some big, burly Marine would be standing there to greet me: "Welcome to Iraq. This is a war zone. Get your stuff and keep your head down."

Well, all I can say is that reality is never quite what I anticipate or what my wild imagination conjures up.

On October 19, the C-130 carrying me, my new best friend Paula, and our computer-whiz buddy Jason landed at Baghdad International Airport. The tail of the plane slowly came down to reveal a brown and dirty runway, with the plane's engines making heat waves on the horizon. We stretched our necks to see anything we could. All we saw was brown stuff and, off in the distance, some buildings that looked like an abandoned airport.

Once off the plane, we were led to some buses. Even though it was October, it was still pretty warm on the tarmac. On the bus, some soldiers directed us to something called a pallet, where we were supposed to get our bags.

Then, the fantasy I'd conjured of my arrival in Iraq appeared right in front of me. As we were stepping off of the bus, I heard a voice boom, "Welcome to Iraq. Get off the bus. Your bags are on the pallet!" But the boisterous instructions weren't coming from the Marine of my imagination. Instead, they came from some pissed-off Air Force chick!

Sergeant Stephanie Clarke was a short, tough-looking woman with a pageboy military haircut, wearing fatigue pants and a body-hugging t-shirt. And her attitude was just as shrill as the toughest Marine drill sergeant—all in her skinny little body! I leaned over to Paula and whispered, "Man, I wonder who pissed her off!" It was almost like a scene from the movie *Stripes,* and this chick was Sergeant Hulka! I could've gritted right back at her because it seemed to me she was looking right at me when she belted out her orders. But then I realized she had a job to do and, just like a drill sergeant, had to be taken seriously. Being a mean chick was just her way of getting it done.

I knew what I was going into before I got to Iraq. This was a war. People were dying every day. We had a serious mission in front of us. And Sergeant Clarke's sober welcome reminded me that this was no game.

I walked past Sergeant "Hulkette" to the pallet to get my stuff. In typical civilian fashion, with no war experience and no one to tell me how to pack, I had a lot of junk. My bag was big and stuffed with everything I thought I would need for the next three years. I could barely hoist it over my shoulder to carry it to the truck. Prickly beads of sweat formed on my body, and they weren't from the warm Arab desert temperatures—they were from pure embarrassment. My bags were too heavy for me to carry. All I needed was this mean chick watching me, laughing at the civilian chick who couldn't carry her shit! I finally managed to drag my stuff over to the truck so it could be hauled to the Green Zone. All the while, I could feel her gaze burning into me, and I swear I could hear her giggling behind my back.

After that drama, Paula and I made our way to the bus that would transport us to the infamous Green Zone, where we would live and work for the remainder of our new lives in the world of Operation Iraqi Freedom (OIF).

Once we got on the bus, I was okay until our host (not the angry Sergeant Hulkette) told us to close the curtains and not open them until we were instructed to do so. We were told that there had been a lot of attacks on a strip of land called BIAP Road, and we had a better chance of not being attacked if no one could see who was in the bus. Now I was scared. I knew I was going into a combat area, but I guess I wasn't quite ready for it to start on the way from the airport.

You know the song "Highway to Hell" by AC/DC? Well, that's what the ride down BIAP Road was like, not just this first time but every time I traveled on it. It had truly earned its reputation. Some months after I arrived in Iraq, a contractor's body was found hanging from one of the bridges on BIAP Road. He was wearing an orange jumpsuit; insurgents had chopped off his head.

It was about an eight-mile trip from the airport to the Green Zone, but one five-mile section was the most dangerous. Even though we were ordered not to open the curtain, I couldn't help it. I had to see what was out there. I pulled the curtain back just a couple of inches with my finger, but I couldn't really see anything, mostly dirt, a few palm trees, and some low-end apartments.

I could see more through the front window. We were on a four-lane road with a median separating the traffic headed to and from the airport, much like something you would see in Arizona or another desert state. Also, I saw a few more palm trees and some other vegetation I didn't recognize. *Vegetation* is a funny word since it brings up images of watered flowers and green plants flourishing in a garden. Nope, that's not what this was. The land was dry and dusty and looked dead. I tried to look for people, but I couldn't really see many from where I was sitting. The story of what would happen if bad guys saw us riding on the bus had me too keyed up to touch the curtain again.

This time, thank God, we survived the 15-minute ride to the Green Zone.

Our bus pulled onto the 2.5-mile-wide compound and stopped in a large parking lot. We were told we'd arrived at the palace. All I could see was a bunch of trees.

Once we were off the bus, I started to get anxious again. I didn't know what I was supposed to do or where I was supposed to go. I had only one piece of information: I was supposed to report to the Green Room. I needed to get to the Green Room in the Green Zone, and nothing that I had seen in Iraq was green at all—just brown and dusty.

Paula was going to work for the governance office. They seemed to have their act together, as someone was there to meet her and get her in the building so she could get situated. No one was there for me. I followed Paula and her guide, explaining my job and where I was supposed to go. It turned out that someone from the office should have met me to get me past the checkpoint and into the palace. Since that person never showed, Paula's guide had to do it. Then I was on my own.

As I walked through the gates, Saddam's palace loomed in front of me. It looked like a cheap version of the U.S. Capitol, with four imposing bronze statues of Saddam Hussein on the top of huge pillars that sectioned off the structure. The lawn was the first green grass I had seen, and it was perfectly manicured. Stretched before the front of the palace was a fountain about the size of a football field, but just like everything else in Iraq, it was completely dried out—an Olympic-sized dust bowl.

I really wanted to take in the grandeur of the setting, but I was anxious and starting to get a little ticked. As I struggled with my oversized bags, I was having an over-the-top selfish moment, just short of a meltdown. *I am here. Where is my greeter? Where's the red carpet? Why isn't someone here to help me?*

When I finally arrived at the Green Room, I discovered it was a huge hall with high ceilings that looked more like a ballroom than an office. It was painted an awful pukey green. Instead of party tables with white linen tablecloths, the hall was filled with junk-cluttered desks. I stood there, taking it in, desperately looking for a knight in shining armor.

Then I saw this guy standing in the middle of this tacky ballroom. I know I must've looked pitiful because he walked over and offered to help—finally, someone to help me! I told him who I was and that I was supposed to report to the Green Room.

"Hi, my name is Chad. I can help you," he said. It was like a voice from heaven. I will never forget that moment and those words for the rest of my life.

Lieutenant Colonel Chad Buehring guided me through the in-processing procedures and stayed by my side until I had found a desk, met my co-workers, and, most importantly, located a place to sleep that night.

He invited me out to dinner with my other colleagues to introduce me to my new way of life. By the time we got back to the palace after dinner, I was exhausted. The trip halfway around the world, the adrenaline overload, and the overwhelming understanding that I was now living in the seat of war had taken their toll.

I went back to my new digs, a white construction trailer with room for four, with two rooms on each side and a bathroom in the middle. I was so physically and emotionally spent that I wondered if I would have the strength to take off my clothes. After three days of traveling, a corkscrew landing on a C-130, a drive down one of the most dangerous roads in the world, and hauling my oversized bags halfway around the world, I finally fell asleep.

As I drifted off, a sudden thought occurred to me: *I wonder who my roommate is?* Just about that time, I heard the trailer's outer door squeak open and then someone fiddling with the lock to my room. Here was the big moment: I would finally get to see my roommate.

When the door opened, the light stayed off. Then I heard an eerily familiar, deep, authoritative female voice. I almost shot up in the bed.

"You left your key in the lock," she said. I almost peed on myself. The only thing I could think was: *Oh, fuck. It's her. The mean girl from the airport!* I was doomed. It was my worst nightmare come true—Sergeant Hulkette was my new roommate. Oh, dear God, how could this have happened?

It turned out Sergeant Stephanie Clarke was a logistics specialist in the Air Force Reserves from Alexandria, VA. Back in Virginia, she was a park police officer. Her job was to help get people and stuff on and off flights coming into and flying out of Iraq.

At some point, we finally had enough time together to talk, just casual conversations about where we were from, our jobs, and other get-to-know-you chitchats.

One morning, I walked into the chow hall in the palace, and she was there, so we ate breakfast together. Steph was not at all the mean chick I had thought she was at BIAP weeks earlier. In fact, she was really very cool and kind. That tough exterior was a thick skin that protected her in the field and on the streets back in D.C. I am so glad I didn't make the mistake of retaliating with my own dirty looks when I first saw her at the airport. Our trailer was too small for two mean girls.

When you enter a war zone, start a new job, or attend a new school, you want to find someone you can connect with. Just like with Paula, Steph and I connected—and boom, that was it, BBFs (bunker buddies forever)! Paula was gregarious and energetic and could make friends with any cold-blooded insurgent. But Steph was such a great roommate because we were alike in many ways.

Steph was kind of a tomboy like me, but she enjoyed being around boys (even though my mom told me never to be around a group of men alone; it makes you look slutty). She would also keep to herself and do her own thing, something I did, too, just kind of shut out the world and get inside my head. She also knew when to talk and when to let silence dominate. Every now and then, when something bad happened, we would talk over the wardrobes or "the wall" in our trailer, just a few words to be there for one another.

One night, I heard a strange, deep thumping sound. I knew what gunfire and explosions sounded like; this was something I'd never heard before. It was the middle of the night, and I didn't know if I needed to brace myself for an attack. Steph must've known I was scared, as she calmly said, "That's us."

"Okay," I replied, relieved that she knew I was nervous but was cool about it and made it seem like it wasn't a big deal.

Friendship with Stephanie also came with a wonderful gift. Since she made frequent trips to the airport, she could get supplies and had access to stuff that other folks in the Green Zone wouldn't see for months, if at all.

One day, I went back to the trailer, and the refrigerator was filled with all kinds of goodies: snacks and drinks, and on top, there was a bottle of alcohol. Alcohol was outlawed by the Iraqis and kept hidden amongst the Americans—that is, until Alcohol Ali showed up in the Green Zone.

Steph was awesome. Living with her was like a scene right out of the TV show *M*A*S*H*. There's always somebody who can get their hands on anything when resources are scarce, and Steph was that person. Once, Stephanie told me we were getting a TV for our trailer. I was thrilled. I hadn't watched TV in weeks and looked forward to something that would keep my mind off the shooting and bombing. Once the TV arrived, I really didn't get to watch it too much; a war zone has a way of keeping you a little too busy for much TV viewing.

As the weeks went by, Steph and I became pretty close. We even developed a little code for times when she needed some time alone in the trailer. I truly came to love Stephanie; she was my calm breeze in the storm of the battlefield.

Then came the moment I dreaded more than our initial meeting: Stephanie received orders to go back home. She didn't want to leave, and I didn't want her to go, either, but it was inevitable. True to her military background, Steph fought the order. She delayed her departure for a few weeks, but by December, she was gone. Her side of the trailer was cold and empty, and for the second time, I felt alone.

In all, I had five roommates during my nearly nine months in Iraq. I should've gone through therapy for that alone. In the end, Air Force Sergeant Stephanie Clark was—and is—the best roommate I've ever had. We don't get to talk much these days, but I know if she called me right now—in the middle of the night—for help, I'd be there for her without question. I love Stephanie. She is and always will be my dear bunker buddy and best roommate.

CHAPTER 4

War Is Hell: When You're Going Through Hell, Keep Going!

> *These two profound quotes describe the horrors of war and how to survive and push through.*
>
> *"When you're going through hell, keep going."*
> *- British Prime Minister Winston Churchill*
>
> *&*
>
> *"War is hell."*
> *- Union General Tecumseh Sherman*

It only took six days in Iraq for me to see that war is hell.

When I walked through the doorway of the Green Room for the first time, I was desperate for someone to help me find my way around and explain what I should do next. I saw a man wearing a Hawaiian T-shirt and khaki pants—a combination that caught my eye from the start.

"Hi, my name is Chad. I can help you," he said.

After Lieutenant Colonel Chad Buehring introduced himself, he took my bags and moved them to the side of the Green Room. Desks piled with dust, papers, bags, and bulletproof vests ran the length of the room. Junk lay in a heap against a glass mirror on the other side of the room. Though no one else was in the large ballroom, it was so full of debris and such a mess that it looked like a combat zone.

Chad led me to a desk over toward the back mirrored wall and said I could leave my things there and work at the desk until whoever it was came back to claim it. That went on a lot for the first few days—desk jockeying. I was an office orphan without a desk, computer, or cell phone. Having a cell phone and a computer was just as important as holding the keys to a vehicle, which was another issue.

Once he found me a temporary desk, Chad helped me get my badge and go through the ordeal of in-processing. In between, he introduced me to my boss, Gary Thatcher, and some of the other people who would be my buddies in the bunker—soon to be known as the IraqPak.

At some point, Paula managed to find me in the Green Room, and she told me she was getting a room at the Rashid Hotel. I hadn't figured out where I was staying yet, but Chad was near and offered to help me with that, too.

We went to a trailer behind the palace, the KBR (Kellogg Brown & Root) office. At the time, KBR was a subsidiary of Halliburton, an oil corporation with very close ties to Vice President Cheney. As a matter of fact, he worked for them before becoming vice president. I really didn't understand all of this before I went to Iraq, but I was reminded of it fairly regularly when news started popping up about KBR allegedly being involved in questionable activities. Initially, my interactions with KBR only included the trailer I lived in, the contractors who cleaned them, and the food and food workers in the chow hall. In short, they managed the living accommodations in the Green Zone and at other locations in Iraq.

When Chad walked me out to the KBR trailer, he mentioned that the Rashid was a good place to stay. He said that most everyone lived there and occasionally went to the little disco club to blow off steam and stress. Two thoughts burned in my brain: One, *There was a club in the war zone?* And two, *Hadn't the Rashid gotten blown up a few weeks ago?* As much as Chad tried to convince me, I just didn't want to stay at the Rashid Hotel, even though Paula was staying there, too.

The KBR lady told me I could stay in one of the trailers behind the palace. I took one more look at Chad before I told her I'd take the trailer.

My new home was a few hundred yards behind Saddam's palace and a few steps away from the palace pool. The path to the trailer went straight back, down a dusty road, along a sidewalk, and past a series of concrete blocks, bunkers used during attacks. Dozens of other plain white trailers and huge palm trees surrounded my trailer. In Saddam's days, without the trailers, it would've been a great place for a pool party or, in the case of the evil dictator, a tree-hanging session for anyone who opposed him. It was beautiful in an Iraqi kind of way. I love palm trees, and I guess that's why I was attracted to the trailer park and wanted it to be my home for the next few months. It was kind of peaceful in an eerie way.

Chad and I arrived at trailer #57-R. A huge palm tree stood right by the door, and a few cloth lounge chairs sat out front. Chad told me that sometimes people would get projectors and watch movies on the side of the trailer walls. I was surprised to hear that people actually watched movies. They had time for that? Well, you gotta have some entertainment, and in a war zone, you had to improvise to get it, I suppose.

The trailer had a bathroom in the middle, with two rooms on each side. I opened the door to my room to see two large armoires positioned in the middle of the room, splitting it in half and making a privacy wall between the two beds. Someone had left their stuff on the side farthest from the door, so I had to take the bed near the door. At least I could make a quick exit if I had to.

Chad left me to get myself situated. I noticed my roommate had an Islamic prayer rug on her side of the room and wondered where I could get one. I started to unpack and set my stuff up so I could at least function. One of the first things I did was decorate my wood-paneled wall with my autographed Bay City Rollers picture (yep, that's another story).

Later that evening, I was back in the Green Room when other people started to shuffle into the office. Chad introduced me and told the group that we should have dinner together to welcome the new chick. When they started talking amongst themselves about where to go, I knew I was in trouble. They decided my induction should start at the Green Zone Café, a little Iraqi restaurant just a few blocks from the palace but still within the 2.5-mile radius of the Green Zone. To further my pledge process, Chad said I could ride with him—and I had to drive.

"Drive?!" I protested. I'd just arrived, I was exhausted, I'd probably fall asleep on my plate, and this man wanted me to drive a vehicle in the dark, in the Middle East, in the middle of a war? Was he crazy? But Chad was right. I needed to get acclimated—and quickly because things happened quickly. Considering the unpredictability of a combat zone, I needed to get my "sea legs" stable as soon as possible.

So, I got into the Suburban, and we drove to the Green Zone Café.

This was my next moment of culture shock: the Green Zone Café used to be a gas station. I could have sworn there were still gas pump slabs right next to the tables.

I had never eaten Middle Eastern food, but at that point, I was too tired and hungry to care. Lesson number two: when you arrive in a combat area, it's inevitable—you're gonna get food poisoning. It's just your body getting used to "some new experiences." I looked over the menu and ordered what I thought would be safe—chicken on a kebab, or *shish taouk*. But it was in a bun and came with French fries. I don't really eat bread or French fries, but I was starving, so I dug into my first meal in Iraq.

After dinner, Chad made me drive back to the palace parking lot. I was so confused and exhausted that I was "punch-drunk."

It would only take a day for me to understand one of the phrases in the Green Zone lexicon—"Saddam's Revenge." The next day, my stomach was cramping, and I had the runs. I had never suffered from stomach issues that bad before, and I was cleaned out! But diarrhea or not, I still had to press on with my job. As it was explained to me, everyone had experienced Saddam's Revenge at some point; you just had to let it run its course and keep moving.

As the days went by, I learned more and more about my job as a press officer for the CPA. My first project was to compile a list of all the efforts to restore and rebuild Iraq thus far. *Time* magazine and *ABC News* were collaborating on a story that would show figures such as how much electricity had been restored to the war-torn country, how many schools had been refurbished or built, how the healthcare system was being restored, and how far along the rebuilding of the oil industry had gotten.

When I started to listen to the background, I knew it was going to be a daunting task to compile everything in the short amount of time I had. From the news stories I had seen before leaving for Iraq, I understood there were problems with rebuilding. At that point, I was stressed out, trying to find out who to go to and where to get the numbers I needed, but I still knew some deep political issues were going on at levels beyond my security clearance.

My first action was to meet with the ABC producer to understand exactly what it was they needed. The producer I worked with on this first project gave me a good idea of what statistics they needed for their story, and I worked in earnest. The network and magazine had been bugging the CPA for the statistics for the story for weeks, and their deadline was quickly approaching. I was the new chick who had to bust ass to get it all together within a few days.

This was a good project for me to start with. It allowed me to get around and meet people who worked as advisors in the different ministries and to understand the issues from the point of view of the people who were working with them. This was an excellent education.

Lesson here: when you get to your theater of operation, make your way around and get to know the people and their jobs. This shouldn't be hard. People love to talk about themselves, and if you do it right, they'll educate you on their job and other things as well. If you can understand each person's role, you can gain an understanding of how the whole body—whether a corporation or a baseball team—works together for the final product or completion of the mission. Get to know people and what they do! It's knowledge.

And here is another one of my favorite little sayings that a very dear friend taught me in college: "Information is knowledge, knowledge is power, and power is the right to self-determination."

I spent the next few days talking to people and gathering information, just like when I was a reporter. I was in my element.

In the meantime, Paula and Chad were my new best buddies. Paula and I were tied at the hip. She would stop by and see if I wanted to go to the bathroom or grab lunch. Chad pushed me to get acclimated to life in the Green Zone and gave me my first ride out into the Red Zone.

On Wednesday, Paula said she had to move. She had been in her room for two days and was already moving to a new spot. She grabbed me, and we went over to the Rashid Hotel to move her stuff from the 11th floor down to the third floor. The Rashid was like everything else that I had seen so far in Iraq—chintzy. Civilians and one or two officers who worked for the Coalition lived there, and the place was pretty much filled up. Journalists had also used the hotel in the days leading up to the war before it was walled off as part of the Green Zone.

After we moved Paula's stuff, she told me her roommate was on leave and asked me if I wanted to stay with her for a few days. Paula was my new best friend, and since I was completely out of sorts, I told her I might like to hang out with her for a couple of days. In the end, though, I changed my mind and decided to stay in my trailer. I felt bad because I loved hanging out with Paula. Her energy was infectious, and I could have used the boost of enthusiasm, but I had to beg off.

By Saturday, I was still feeling like crap. Saddam's Revenge had completely cleaned out my system, and Mother Nature had decided to do her frontal assault on my lower abdomen. I was miserable.

Next thing I knew, Chad walked over to my desk, kneeled down, and tried his best to convince me to go to a party at the Rashid. I told him I loved him to death, but I was feeling like crap and was going back to my little trailer to get some rest. He finally eased up and told me he hoped I felt better and to have a good night.

The next morning, Sunday, I woke up feeling bad for turning down Paula and Chad, so I decided to walk over to the Rashid. Now, since the Rashid was at least a mile away, this was more of an exercise routine than a quick walk. I had just started out when, all of a sudden, I heard *boom, boom, boom, boom, boom, boom*—six explosions, and they sounded very close. I was standing there, startled, when my friend Jared came running up. It looked like he had just rolled out of bed and thrown on whatever was on the floor.

"What was that?" I nervously asked him.

"It sounds like a bombing, and it sounds close," he said. The first thing he did was pull out his phone to call Chad. Chad was the office manager and the office rock; if we needed anything, our first call was to Chad. He'd fix whatever was wrong.

I didn't think anything of it when Jared said Chad wasn't answering and hung up his phone. As Jared and I walked into the Green Room, things were starting to get chaotic. That's when we learned mortars had hit the Rashid. There were injuries, and apparently, the situation was a mess. We decided that it was best not to go over there because we didn't know if more attacks were on the way or what was going to happen. Meanwhile, Jared kept trying to call Chad, but still no answer.

As the morning turned into late morning, more information started to trickle in about the attack. Deputy Secretary of Defense Paul Wolfowitz was in Iraq and happened to be in the Rashid during the attack.

By now, everyone was in the office, and the place was abuzz. Then the news came. Someone told us that Chad had been seriously hurt in the bombing. I was shell-shocked. Chad was hurt? I had just seen him the night before, kneeling at my desk, and everything was fine!

My first thought was that I should go down to the Combat Air Support Hospital (CASH) to see him. When you're not trained in combat, as I wasn't, you don't think clearly. All I could think was that I needed to go see if Chad was okay. I didn't say a word to anyone; I just quietly walked out of the Green Room, out the front door of the palace, and down the road toward the CASH. I was numb. We'd just been attacked; I'd never heard a noise like that before, and I had felt my chest thump with the power of the blasts. My system was overloaded. My only thought was to get to the CASH, and I kept walking.

I was about halfway there when I saw another one of my Green Room colleagues, a friend of Chad's, too, and since we had only met a few days before, I couldn't remember her name. She was on the same mission I was. She told me her friend Chad had been hurt in the bombing, and she was going to the CASH, and then she asked me to go with her.

When we got to the CASH, we were turned away, and I went back to the Green Room. As I was walking up the driveway to the palace doors, my friend Shane came out and over to me. He was distraught, his face was red, and it looked like he had been crying. I already knew what he was going to say.

"Chad's dead."

Shane grabbed me, and we hugged. I walked into the palace with a supreme sense of heaviness, like I was carrying a coffin on my back.

When I walked into the Green Room, everyone was either crying or had just wiped away the tears. My boss, Gary Thatcher, approached me to seal in the bad news. "Chad is dead," he said, and I told him that I knew. Gary and Chad were just as close as Paula and I.

Days later, at Chad's memorial, Gary told us that the morning Chad died, he had just signed the papers for Chad to spend another tour of duty in Iraq. I've known Gary for several years now, and I can honestly say he took Chad's death harder than the rest of us. Years later, every time I saw him, he still had this sadness about him, like he had lost his best friend.

Moments after I learned of Chad's murder, I thought, *Well, where the hell is Paula?* I knew she should've gotten in touch with me by now to tell me she was okay. I told Jared I hadn't heard from Paula, who was also at the Rashid, and asked him if he could help me find out if she was hurt. Jared suggested we walk over to see Ambassador Pat Kelly, as the ambassador had a list of all those injured in the bombing and might know about Paula.

We walked across the rotunda and through the double doors into the executive offices. Jared introduced me to Ambassador Kelly and told him I was trying to locate my friend. Just then, the doors opened, and a petite blonde woman interrupted us. "Is your name Traci?" she asked. I said yes, and all I heard her say was, "Paula sent me to get you. It was a clean break."

"What?" I said, and she repeated herself. Paula had been hurt in the bombing and broken her arm, but it was a clean break. I thought my knees were going to buckle. I had just learned that my dear friend Chad was dead, and now Paula was hurt. "Paula? Not Paula!" I screamed. Tears rolled down my face, following the tracks where the others had just dried.

Ambassador Kelly grabbed a tissue and let me have my moment. When I was finished, he said the smartest thing anyone could have said, "Now, get back out into the fight." I thanked him and headed back down to the CASH for the second time that day.

On the way down, Beth Payne (the U.S. consulate in Iraq) filled me in on what had happened to Paula. It was worse than a simple broken arm. A mortar had slammed into the room and hit a chunk of concrete. The concrete had slammed into her and nearly torn off her right arm. Because there weren't a lot of ambulances and the hotel was still under attack, Paula had to rush down three floors with Beth applying pressure to the wound—not a tourniquet. By not tying a tourniquet around Paula's arm, Beth had saved it. Paula later told me that after this incident, the State Department made changes to the way they taught their safety classes, including adding instructions on not using tourniquets.

Before leaving for Iraq, I had told myself over and over again that I would see blood and even death. I didn't go through the formal training soldiers do, so I had to find a way to mentally prepare myself for the war zone on my own. Even though I had been briefed on Paula's condition, I still wasn't prepared for what I saw.

When I walked into the room, Paula was lying there with her right arm up in this sling contraption and these huge gray pins that looked like oversized knitting needles sticking out of it. The pins were holding her arm together. A bandage was wrapped around her arm, so thick that it looked like a cast. Despite its thickness, blood was still seeping through.

I was so happy to see she was alive, and I was determined not to cry in front of her. She was sedated and in a lot of pain, but her bubbly personality still managed to shine through.

I wanted to give her a hug, but that was impossible, and quite honestly, I was afraid to go near her. She started to talk as best she could under sedation. She told me she had been in the bed against the wall across from the window. Next thing she knew, she felt this terrible pain in her arm and thought the mortar had torn her arm off. She asked me if anybody had been killed, and I had to tell her that Chad was dead.

As we talked, the medication started to wear off, and the pain pushed through. "Talk to me, talk to me," Paula begged. "Talk to me so I won't feel the pain." When Paula and I got going, we could talk for hours. But at that moment, when she needed me, I couldn't think of anything to say. My mind went blank. What do you say to your battle-zone best friend after the worst moment in her life, as she is lying there with her arm nearly shredded off?

I just started talking; I told her I wanted to write a book about the First Lady of the United States. I didn't know if I was making sense. I could hear what I wanted to say in my brain, but when the words came out of my mouth, they didn't seem to make any sense at all. I left to give Paula some rest, but I vowed to her that I would be back.

When I returned to her room later that evening, Deputy Secretary Wolfowitz was there. He thanked her for her sacrifice and gave her a coin. Coins are kind of a big deal in the military. Highly ranked officers and civilians have special coins made up, kind of like a calling card to hand out to special people on special occasions. Soldiers collect them like baseball cards; the higher the rank, the better. They let others know that you met with powerful people. There's even a special way to hand them out. The giver extends his hand as though to shake the receiver's hand, and the coin is discreetly placed in the receiver's palm. Even though it was an honor for Paula to get the deputy secretary's coin, the circumstances were just not worth it. Paula couldn't even hold it because of her arm, so I had to take it for her.

I wondered what Paula's room looked like after the attack, and a couple of days later, I found out. Paula asked Beth and me to go to her room and grab a couple of things she needed right away. Her injury was severe enough that she was going to have to go to the military hospital in Landstuhl, Germany, and then to Walter Reed Army Medical Center in Washington, D.C.

As Beth and I went to the Rashid, I was once again in the position of trying to brace myself for what I might see but not really knowing what to expect. Paula had talked about a pungent smell after the attack, and I waited for it to fill my nostrils when we first walked into her room.

But the smell wasn't what struck me. No, it was the general mess that the room was in; it looked like a bomb had blown up in it. There were two beds, one along the wall, perpendicular to the window, and one across from the window. Dust, debris, and scattered items lay everywhere. Chunks of concrete covered the bed near the window, and I realized that was where I would've been sleeping if I had stayed with Paula.

Beth and I worked quickly to locate the items Paula needed: her computer, some clothes, her birth control pills, and toiletries. I went to grab her computer, and that's when I saw it: a chunk of Paula's flesh on the floor. That's when it happened. That's when I learned to shut off my emotions completely. I didn't even feel numb. You can at least feel a slight buzz when you are numb. I felt nothing at all—just black. I would learn to go back to this place whenever I had to push through something.

Shutting everything else out, I picked up the computer, and we were out of there. It's funny—as much as I was waiting for the terrible smell Paula had told me about, I never did smell it. Does shutting down emotionally affect your ability to smell?

When Beth and I went to drop off Paula's stuff, I didn't have the heart to tell Paula everything we had seen. She didn't need to know.

Once Paula was medevaced out of Iraq, I called around and tracked her down to Walter Reed. I was so relieved to hear her voice. In all, Paula had 14 surgeries to clean out her wound and repair the damage. But as with so many soldiers wounded in battle, Paula also suffered emotional wounds that have taken many years to heal.

A couple of days after the attack, we had a memorial service for Chad. It was hard. Soldiers have a special ceremony, the Final Roll Call, which is a missing-man formation. A small memorial was placed in the front of the chapel, along with his boots, gun, dog tags and an American flag. Chad's unit gathered together, and the roll was called. When a soldier's name is called, he answers, "Here," but when the dead soldier's name is called, it's silent. It's a very powerful moment.

That's when Chad's death hit me hard. I cried my eyes out, staring at his stuff and hating the war. After the service, we had a group therapy moment. The casualty assistance officer had us sit down to talk about "it." We had to go around the room and share our thoughts and anything we wanted to say about Chad or the attack. Our Iraqi receptionist said something that sticks with me today. She said that in Iraq, they have a saying for those who pass away: "God needed that person for another beautiful flower in His garden." That thought was soothing to my soul.

After six days, I'd lost Chad and seen Paula in so much pain that I could've left Iraq. I could've gone back home and stayed safe in D.C., but this is where Winston Churchill's famous quote comes in: "When you're going through hell, keep going." To those out there reading this, no matter how bad it gets, keep going. Take one breath at a time, one step at a time, one day at a time.

In this case, I wasn't prepared to sit and talk with someone one minute only to be told they had been killed the next. But despite the dangers and the death, I didn't leave Iraq. I wanted to see how things turned out, and I didn't just want to give up. Churchill has another famous quote on that, too: "Never, ever, ever give up."

Weeks later, Chad was buried at Arlington Cemetery. At the time, he was the highest-ranking officer killed in battle. I called my mom and told her about Chad and how kind he had been to me those six days. Mom was so sweet, and she and my PopPop went to Chad's funeral at Arlington.

On the day of the service, my mom called with some wild news. She was at the cemetery, and she told me I wouldn't believe where they had buried Chad. He was right near Daddy's grave, across the way and a few steps down. Daddy had died in 2000, and Chad in 2003. I hadn't expected their deaths to be so close.

Now, seven years after Chad's death, Mom is buried near his grave, too.

War is hell. I just keep moving forward.

CHAPTER 5

My Friend Fahmi

> *Here are a few key lessons for this chapter: 1) When visiting a foreign country, learn about the locals and see things through their eyes; 2) Define yourself before someone else has the chance to define you. Once a perception is created, it's difficult to redefine yourself; 3) Don't trust anyone!*

"Sweet Source of Trouble." That's what my friend Fahmi used to call me. I guess I earned the name since I nearly got him killed.

Fahmi and I met in November of 2003. As special projects coordinator for the CPA, my mission at the time was to direct the press coverage of the removal of the infamous Saddam Hussein statues that were perched above the presidential palace like bronze buzzards. Fahmi was the Iraqi contractor chosen to remove the gaudy statues we referred to as the "Saddam Heads."

The Saddam Heads stood as a gothic reminder of Saddam's megalomaniac reign. Their removal symbolized his removal from power and the hope for a free, democratic Iraq. Plus, the things were just hideous.

The CPA chose Fahmi to bring down the Saddam Heads because he had assured them, with effusive confidence, that he could bring down the 40-foot bronze statues without so much as a scratch. Also, it was more appropriate for an Iraqi to bring down the statues and avoid the scene that had occurred at the beginning of the war when, in Baghdad's Firdos Square, U.S. soldiers had played a large role in bringing down the Saddam statue.

Fahmi is one of the most endearing people I know. From the moment we met, it was as though we had known each other for years. Fahmi was a diminutive Iraqi with gray-speckled hair, somewhere in his fifties. He was gregarious, energetic, kind, and humorous, with an astounding level of pride and confidence that was just this side of arrogance.

Our friendship started with us discussing his ability as an engineer to bring down the Saddam Heads, but as we grew more comfortable, the conversations quickly turned into shouting matches about the best way to do the job. I learned before I traveled to Iraq that "passionate discussions" were a culturally added value of the Middle East. Seeing two men arguing in a market over the price of an item was a common sight, and Fahmi, true to his values, would describe with great passion his ideas and plans for the correct removal of the statues.

Our discussions soon led to talk about the war in Iraq and whether or not the Iraqis would be able to govern themselves. We never gently touched on a subject to feel out whether it was safe to discuss. No, Fahmi was bold enough to jump right into what was going right or wrong with the rebuilding of his country, and he was always painfully honest. If the electricity were not staying on long enough, he would say so. He helped me understand the true state of things in Iraq.

More than being honest, Fahmi was always hopeful about the success of Iraq. When things seemed particularly bad, he would come up with some nugget of wisdom. "Traci," he would say, "a woman 'with child' will wait nine months to grow inside her, and then she will go through great labor pains to bring a child into the world, but once the baby is born, all the pain she has gone through is worth it. Look, a beautiful baby is born." That is how Fahmi used to describe the rebuilding of Iraq, as a woman in labor. After great pain, a wonderful child would be born. I was supposed to be helping Fahmi, but in all truthfulness, he was helping me.

For this special project, I played the role of negotiator between three parties: Fahmi, the engineer; the military officers in charge of the premises; and the media. The military and the Green Zone security team wanted me to assure them that inner security would not be compromised, and the media wanted to broadcast around the world a historic picture of the heads coming down. At the time, members of the Western press corps were upset to learn that they would be severely restricted in their ability to take pictures. But in this case, I had to be stern with them.

Before cranes could be put into place, Fahmi and his workers started receiving death threats. You have to understand something about Iraq. When someone was threatened, there was a very good chance the person or party who made the threat would follow through. We lost countless Iraqis to assassinations—mayors, TV reporters, and others who dared to cooperate with the Coalition.

Since Fahmi's life had been threatened, I was forced to prohibit the press from taking pictures of him and his team. To protect Fahmi's identity and still give the media what they needed, I had two choices: invite the entire press corps, Pan-Arab and Western, to view the removal from outside the grounds of the palace, or organize a small press pool and allow them to take pictures and videotape from inside the palace, which would provide them with the best images.

Eventually, I chose to throw together the press pool and take them up to the roof so they could have the best vantage point, sitting by one of the Saddam Heads while another one was being removed. However, I had to warn the press before, during, and after the heads came down that they were not allowed to show close-up shots of the workers. If they did, I would confiscate and destroy their film or tapes.

During the days leading up to the "big day," Fahmi and I forged a friendship, growing so close that we got into passionate discussions—more like heated arguments—about which head would be removed first, where the media should be placed for the best view, and which angle the media might want to shoot from so they couldn't get a sharp image of Fahmi. In the end, Fahmi came to trust me and my judgment. He would say in his thick Arabic accent, "Traci, you promise. No pictures of me and my workers!"

And I would always respond, "Yes, Fahmi, I promise. No pictures." I meant it, too. My word was my bond.

The removal of the heads was often delayed because of equipment and technical problems, or the workers had problems getting through the checkpoints to the palace. But Fahmi was more often delayed because of a thing we outsiders came to call "*inshallah*" time. *Inshallah* is Arabic for "God willing," so this was "God's time"—or, as we joked, "whenever." Fahmi was never on Western time. He was always on *inshallah* time. If he said 11:00 AM, you wouldn't see him until 1:00 or 2:00, but eventually, he would show up.

After a couple of weeks, the day finally came for the heads to come down—or, as I liked to say, the day arrived for the "heads to roll." Fahmi was there with his workers and cranes, ready to go. The press pool, consisting of a print reporter, print photographer, and TV crew, was in place as well. We walked around the back of the palace, up a back stairwell, onto the roof, and over a couple of walls to get right next to one of the four Saddam Heads. A crowd of CPA staff, Iraqis, and other Coalition folks had gathered outside to witness the historic occasion.

As the crane crept over to the third head, I got a call from one of my IraqPak friends. He told me the craziest thing. On this partly cloudy day, a rainbow had appeared right over the Saddam Head we were standing under. But there wasn't any rain!

A few moments later, the wires were attached to the bronze statue, and the crane lifted it up and gently guided it to the ground. The crowd below cheered the moment the head hit the ground. Fahmi had succeeded. The first head had been toppled without any gunfire or flying mortars. The moment had been recorded in history, and I was proud to play a tiny role in it.

The next day, I received an urgent phone call from a frantic Major Phil Hicks, who worked with me on the project. It appeared an Iraqi TV network was on the grounds videotaping the workers as they were bringing down the second statue—without my permission or knowledge. I ran outside and told the crew to hand over the videotape. I took the tape to the TV station and talked with one of the producers, who was American, about the violation. He assured me they would not broadcast any images that might be used to identify the workers. I took him at his word and handed him the videotape.

The next morning, I received another frantic phone call. This time, it was Fahmi. He was yelling and so angry I couldn't understand what he was saying. The only phrase that came through loud and clear was, "Traci, you promised me! You broke your promise!" I immediately ran out of the palace to find Fahmi. I found him standing on the front grounds, enraged. He told me that some friends of the workers had called them to say they had seen them on TV helping to bring down the statue. I was furious! Worse, Fahmi and his workers were now in fear for their lives.

All I could say to Fahmi was that I was very sorry and I would do what I could to fix it. I then told him all we could do was pray that no one was targeted. Fahmi yelled back at me, "Pray? I pray five times a day!"

I rushed over to the TV network with the weight of Fahmi's words weighing heavily on my shoulders and my heart. Fahmi had become a dear friend to me, and now his life was in danger because of me. I was devastated. Why had I trusted the producers?

When I arrived at the station, I went right over to the producer and told him they had violated the rules and would face severe consequences. I demanded that they give me the tape immediately. I took the tape, went back to my supervisor, and told him the severity of the situation. The next few days were horrible; I was on edge. I couldn't sleep because I was so worried about Fahmi and the workers.

A few days later, Major Hicks called me again, but this time, he wasn't frantic. He said Fahmi and his team were having a picnic lunch on the palace lawn and wanted me to join them. I was happy to accept. They had everything but a red and white picnic blanket. As we sat and ate, Fahmi interpreted. We laughed a lot. Someone would say something, and even though I didn't understand, I would laugh along with everyone else. But a nagging voice in the back of my mind reminded me that any of these men might meet his death because of my mistake.

As the days and weeks went by, Fahmi and I grew closer. Because he had successfully removed the Saddam Heads, he received more contract work. Some days, he would call to say he was on his way to the Green Zone and ask me to go to lunch. We would talk like long-lost friends. I learned about his family, his wife, Maha, and his three daughters, Mayse, Teba, and Deana. He also talked about what life was like when Saddam was in power and how everyone feared his unpredictability. He even shared with me what happened to him the night of "shock and awe." We remember the images of the fireballs as they scattered across the city of Baghdad, but I had given little thought to the personal stories of Iraqis who had survived the invasion.

The attack seemed to go on forever. Fahmi described how, after the bombing started, he drove across Baghdad to get to his family. He was behind a truck that got hit with something, and the explosion shattered his windshield. The car was still drivable, so he forged ahead. He also described his horror when he felt pain in his legs and looked down to see his blue jeans had turned a darker shade. Shards of glass had cut his legs. He was in a great deal of pain, but his adrenaline was pumping, and he did everything he could to get to his family. He eventually made it home, and his family made it through the attack unscathed.

By the spring of 2004, Fahmi had told his family so much about me that it was inevitable that we would finally meet. But that was another issue. Security was unstable in Baghdad, and we had been told on many occasions that we could not leave the Green Zone without permission and security. This was one time I felt like breaking the rules was worth the risk. Fahmi met me in the parking lot and handed me a shawl to cover my head. Let's just say, as a Black woman, even though my Iraqi brothers and sisters had a tawny complexion, I was still too dark-skinned not to get noticed. As I was putting the shawl over my head in Fahmi's car, he told me that if things got uneasy, I would have to hide under the dashboard. When we pulled out of the Green Zone, I edged toward the front of the seat, ready to duck down in case anyone spotted me.

Moments later, we made it safely to Fahmi's house. As we drove up to the bi-level, I noticed the neighborhood. The first thing that stuck out was the number of generators. Every house had one. Since Baghdad's electrical system was not sufficient, most of the city was only getting around eight hours of power a day—on a good day.

Fahmi's family greeted me like the prodigal child returning home. Maha, Deana, and Teba all gave me the biggest hugs. Unfortunately, Mayse was not there, and I never got to meet her. They were living in the upstairs portion of the two-level house, and Fahmi's parents lived downstairs. Maha had cooked a lavish dinner, so much food I couldn't eat everything—the Iraqi way.

Before dinner, I had a chance to look around their modest home and was surprised to find that it could've belonged in the States. To my surprise, a Brittany Spears video was playing on the TV! Teba and Deana sang along. It was such a funny moment. I know the girls didn't understand the meaning of each word, but they belted out the song as if English were their native language.

Toward the end of the evening, I pulled out the gifts I had bought for Fahmi and his girls—mostly candy and some small toys, but it didn't need to be much. The gift of being welcomed into the family was priceless. I left Fahmi's house feeling like I was leaving my Iraqi home. In fact, the girls were calling me Auntie Traci when I left.

My final departure from Iraq was bittersweet. For security reasons, I had to keep my departure a secret. I did not get to say a final goodbye to my friend Fahmi.

After I returned to the States, I received an email from Fahmi that said, "I stopped by the Green Room to say hi to my friend Traci, and you were not there. Please tell me you have not left without saying goodbye." I was heartbroken. I had to write Fahmi back and tell him the reason for my disappearance.

Despite that, for many months, I received emails from Fahmi addressed as *"Sweet Source of Trouble."* But as I sit and write this, it has been several years since I last heard from him. His last note did offer me some comfort. He said he had moved his family out of Baghdad because an insurgent had tried to kidnap one of the girls. Thank God she had managed to escape. The last I heard from Fahmi, they were living in Jordan.

You may be wondering why I haven't mentioned the family's last name. It's not because I have forgotten it. As you can imagine, after more than six years, I still am quite protective of my friend Fahmi.

CHAPTER 6

BBFs (Bunker Buddies Forever) Are Like BFFs: You Gotta Have Good Girlfriends to Get Through a War

> *When you're in a war zone, it's great to have a good girlfriend by your side to get through the rough days; it just makes life a little easier. But in a war zone, you may develop a friendship that goes beyond a BFF. It's a friendship that no one ever told me about and only those in a war zone can completely understand.*

By December 2003, I had lost two of my bunker buddies. Paula's arm had nearly been blown off during the attack on the Rashid Hotel, and she had been transported back to the U.S. to heal. My roommate, Steph, had reached the end of her duty and transferred out of Baghdad, too. I was left to play in the Iraqi sandbox alone. No, I wasn't completely alone, but losing Paula and Steph were blows.

Let me explain the concept of a bunker buddy. If you've ever seen any World War II film footage, the old black and white stuff, soldiers are down in a foxhole together, shivering, shooting, or getting shot at but still in the fight. They're down in the trenches, looking beat, smoking cigarettes, and waiting for the fighting to start or end. In Iraq, the security folks built concrete structures called bunkers that looked like small carports. They didn't have doors or anything, just a square block that provided protection during an attack. When rockets started flying, you'd run and duck for cover in the bunker and share the small space with a bunch of folks, thus the term "bunker buddies."

Let me explain a little more. I am not old enough to remember Vietnam, but if you've watched the History Channel, you've seen the images of soldiers wading together through rice fields, waist-deep in water, looking for the enemy or waiting for an attack. There's something about wading through a war zone, with your buddies to your left and right or covering your back, that brings you closer together. No one can understand it unless they've been to war.

In the military, they have the term "wingman." Imagine a clock: while you're looking up at twelve, your wingman's watching your back, or your six. Wingman also refers to the pilot of the plane flying just off the lead plane's wing. You've probably heard of the term used in the dating world, too. A wingman is a guy's buddy at the club who bails him out talking to a "waste of time." I think that paints the picture.

Around November, one of my IraqPak colleagues said a "new girl" was on the way to the Green Zone, and they asked me to keep an eye on her when she arrived. I took this request seriously. The combat zone was no joke, and if she went through what I had in my first few days, I knew it would be tough for her. Me watching out for the new girl was not going to be an episode of *Mad Men*, where the "experienced" secretary shows the ropes to the new secretary with a hint of cattiness. I intended to truly watch out for her and treat her just as well as my two friends had treated me.

Kristi Clemens arrived in the Green Zone in December of '03, and we hit it off right off the bat. The more we were around each other, the more we realized we mirrored one another. Kristi was beautiful, stylish, and tough; she was everything I aspired to be, minus any hint of jealousy.

Kristi was also spontaneous. I'd say, "Hey, let's walk over to the Rashid Hotel and look at some rugs." Without hesitation, Kristi would jump up, and we were out the door. We'd walk into a shop and head over to the same rug to buy it. Kristi was my new BBF, my bunker buddy forever, kind of like a BFF.

One night, our guys were dropping some packages on our insurgent friends, and it was getting pretty loud. I knew Kristi would be anxious, just like I had been my first few nights in the sandbox, so I called her and told her, "Hey, that's us." I could tell she was relieved not only because we were not under attack but because somebody had called to ease her worry.

Just like with Paula, Kristi played trailer-park roulette and had to move from room to room in the beginning. Once again, I was in the position of helping my BBF pack up her stuff and haul it by hand from one part of the trailer park to another. But this time, Kristi's room didn't get hit with a barrage of mortars like Paula's had.

Not too long after Kristi's arrival, Susan Phalen showed up in the Green Zone. Susan was tall, with dark hair and delicate features, and was quite the opposite of Kristi—real laid-back and unexcitable. While Kristi had all cylinders firing at full throttle, Susan ran in neutral, slow to anger or get emotional, cruising through whatever came along.

Susan and Kristi took up where Paula and Stephanie had left off. When we were under attack and had to run down to the shelter, we always looked around for one another to make sure all were okay. Even though we sat right next to each other in the office, we ate lunch and dinner together, helped one another with projects, and even traveled together.

I look at some of these reality shows on TV today, where girls backstab, punch, fight, and kick one another over some boy. Susan, Kristi, and I would never do that to one another. Real BBFs don't. We all love one another, and playing the backstabbing "mean girl" role never enters our thought processes.

Here's the best way to describe a bunker buddy.

Let's say you're at home one night, relaxing, watching TV, and your BFF calls. She's on her way to the mall to find some shoes for a party she's going to, and she needs you to go with her for a second opinion. You're lying on your "crack couch," thinking, *I don't want to get up, get dressed, and drive down 95 to the mall. I am not feeling it.* So, you tell her you are in the middle of something else… (blah, blah, blah), and you stay there in your comfort zone. But if your BBF calls and says she's on the way down to the CASH to drop off some magazines and candy to wounded soldiers, you're like, "I'm there. See you in a few." When you're in the hot spot with your bunker buddy, you can't leave her hangin' out there alone. You don't want to let her or the mission down.

Here's another example. On Christmas Eve of 2003, Kristi and I were going to a party in the Green Zone. Christmas was a dangerous time for the Coalition. Insurgents and terrorists knew the importance of the religious holidays for Christians and exploited them as a prime time to launch an attack. So, with this in mind, we had to be careful about where we went.

As we walked, Kristi abruptly stopped, shouted an expletive, and said she'd lost her scarf. She said it was her favorite scarf and that it carried sentimental value. We looked around in the dark and didn't see it right away. After a few moments, she said to forget about it. I told her no, that we were going to search around until we found it.

I had a good reason for helping her. You have so little in the theater of war that you treasure what you do have. You live in a trailer, a tent, or out of a bunk bed in a huge ballroom, so you'll bring some little trinket or something special that'll fit in your rucksack to remind you of home or your family. When I left to go to Iraq, I asked one of the guys who had already been there what I should take with me, and he told me to take something I'd miss. Kristi had brought her scarf. So, I understood what she was feeling. She had lost something she treasured. More than that, Kristi was my bunker buddy, and if she needed my help, I was willing to walk around until daybreak to help her find that treasured item.

So, while Kristi poked around in the dark, I retraced our steps up the street. About halfway down the block, I saw a dark clump on the sidewalk—it was Kristi's scarf! I picked it up and yelled, "Kristi, I found it. I found it!" You would've thought I'd found Saddam Hussein! I was so happy! Kristi was surprised that I had been so determined, but that's what BBFs do. They go beyond the extra mile (or block) to help their buddy.

I am willing to take this a step further. In a war, sometimes you have to be stronger than you ever thought possible. When I share my experiences in Iraq, some people say, "I could never do that…" (go to a war zone, get shot at, whatever). Let me clue you in on something quite critical here: YOU have untapped potential in a battle zone, whether it's in Iraq, Afghanistan, working the cash register at Walmart, or wherever. You have the potential to be stronger than you ever imagined.

If someone had told me that within the first six days I was in Iraq, I would have to go into my dear friend's room, walk through the dust and mess, look down to see a chunk of her arm on the floor, and have to stay there and work around it, trying to pretend I hadn't seen it, I would've said, "Hell, no! I can skip that!" But I had to reach down into a place where I hadn't been and pull something out of me that I didn't know existed. I went there because Paula needed me to be strong for her. She needed me to focus for her—I didn't have a choice.

Finally, I will take this even deeper. That place I had to go that allowed me to cut off emotionally when I was in Paula's room also gave me something else, something that only BBFs share: that dark place… the black hole. Your BBFs know this place because you all go there from time to time after undergoing dramatic, violent experiences. You go there instinctually, but you shouldn't stay there for too long because, if you do, you may not be able to come out. I've seen soldiers who've been in a combat zone several times and can't quite come out of it. They have this lost, blank stare like they are in a dark cave and can't get out—they're just gone. Your bunker buddy may never verbally admit that she's been to this place, but she doesn't have to. You've been there with her.

After we all left Iraq, I was diagnosed with a tumor in my breast. Even though we had gone our separate ways, working and living normal lives again, Kristi left work and came to the hospital and sat with my mom during my surgery. A year after that, when my mom died of cancer, Kristi and Susan showed up at the funeral and held my hand during the ordeal. Susan has suffered some drama of her own; we both have "bubbly" stomachs because of our days in Iraq. We call each other and share symptoms and details that are way too icky for public consumption.

On the upside, Kristi recently got married and opted for a small wedding. They had a guest list of two hundred that they cut down to 50. Who was sitting in the second row? Her BBFs: Susan and me.

That's a BBF. You can have blood sisters, sorority sisters, and play sisters, but there's nothing like a BBF. She stays with you in the trenches, tells you the truth when you don't want to hear it, lends you her shoulder when you need it, and feels your deepest pain since she's been in the bunker, too.

CHAPTER 7

Give an Iraqi Boy a Toy and Pray He Doesn't Trade It for a Rocket Launcher

> *"A child's life is like a piece of paper on which every person leaves a mark."*
> *—Chinese proverb*

I have a special little place in my heart for a group of Iraqi boys. By now, they've grown into young men and possibly insurgents. I hope with all my heart that the bit of time we spent together achieved the "winning the hearts and minds" mission we heard so much about while we were in Iraq. I hoped then, as I do now, that those young men would think twice before accepting a few bucks to lob mortars at the "Americans."

It was the fall of 2003, football season in the U.S., and just because we were thousands of miles away in the desert didn't mean soldiers wouldn't take an opportunity to enjoy the fall rite.

One of the soldiers mentioned that a unit had partnered with an orphanage in a nearby district, and they were going to videotape the Iraqi boys yelling a cheer for one of the college teams. He mentioned they might need help and asked if I wouldn't mind going to help record the greeting. I had only been in Iraq for a few weeks and really wanted to get out and meet some of the Iraqis, especially if they were children, so I went.

Jumping into a car to head over to an orphanage across town sounds simple enough, but in a war zone, nothing is ever easy—or safe. First, I had to let someone from the office know where I was going, just in case I didn't make it back. Then, I met up with the soldiers in the parking lot of the palace and went through the ritual of getting ready to enter the Red Zone. This meant putting on a 30-pound flack vest.

Trying to get into that thing properly was like using just your thumbs to put on your bra. First, the body armor was heavy as hell. You had to pick it up, hoist it over your head, and slide it down over your chest. I would try to be cool and show I was strong enough to lift it over my head without any assistance—which usually meant I struggled long enough for a soldier, feeling sorry for me, to step over and help me heave it over my head. Then I had to secure it down with Velcro straps, which felt like pressing a car hood over my boobs. It was not the most comfortable garment, but I knew it might one day save my life.

After all this prepping, we were finally ready to jump into the Suburban and head over to the al Waziriya Orphanage. This was only my second trip driving out of the Green Zone, and even though it was broad daylight, the drive was just as crazy as my first. Of course, someone was riding "shotgun," looking out for any suspicious characters or vehicles that might try to ram us and start shooting.

Oh, I forgot to mention that at this time in Iraq, there were no rules for driving—none! We were in a big Suburban and did not have a security team to escort us to the orphanage. We were on our own. I must admit, I enjoyed it! The traffic was insane, not because there were lots of cars on the streets but because nobody paid attention to the streetlights—if there were any that worked. We drove through one intersection where there was an Iraqi traffic cop. Even though he was a real cop, waving his arms and trying to direct traffic, nobody paid attention to him. It was almost like he was a street decoration.

We drove down one way, couldn't get through there, and then decided to use the median as a turn lane to go back in the other direction. The Suburban jumped and bounced like we were driving through a pock-marked obstacle course.

We finally made it to the al Waziriya Orphanage. I honestly can't say what I expected to see. It smelled of fresh paint and was quite clean and neat. The soldiers had renovated the cement building and painted it so the boys and girls could have a decent place to live. In Islamic culture, if children lose their parents, they go to live with another relative. If one is not available, like American children, they end up in an orphanage. Also, in Islamic culture, the boys live separately from the girls. For this trip, we were going to visit the boys.

One officer was from Louisiana, and he wanted the boys to shout a cheer for LSU and their upcoming football game. During previous trips, the soldiers had taught the Arabic-speaking children to shout something like, "Go, LSU! Go Tigers!" Another officer had given them a big purple and gold banner with the school's logo on it to hold.

We lined the boys up, and the soldiers helped them practice their cheer. The boys were so well behaved and followed every order given to them. They paid attention and seemed to respect the soldiers' authority. To add to the ambiance, one soldier brought along a football. Since the boys didn't really have any toys, they were absolutely fascinated by the football. No, I am not talking about a black-and-white soccer ball. I am talking about the real deal, the true American icon, the pigskin.

After the soldiers took their pictures, the boys were milling around, and I decided to teach them how to throw the football. Finally, my tomboy days came in handy! I picked up the football, showed it to the boys, and said, "Football." I repeated the word several times and then decided to show them how it worked.

Now, I know the soldiers had been there before, and I am not sure if they played football with them, but this seemed a new concept for them. Before long, we were on the playground, tossing the football back and forth. Even though there was a language barrier, the laughing and playing didn't need interpretation—we were having a blast!

Soon—too soon—it was time for us to get back in the Suburban and head back to the war. I wasn't sure when I would return to the orphanage, but I asked the soldiers to please allow me to come back with them the next time they paid a visit.

It wasn't too long before we were headed back to the al Waziriya District to visit my new friends. I was really looking forward to seeing the boys and hoped they would remember me. This time, not only did we have a football, but we had boxes of toys for the boys to play with. Somehow, most of the toys were pencils, books, and dolls! Yes, I said dolls for boys to play with.

Now, you and I both know that this would not fly in the States. A boy down the street playing with Barbie would be written off as "effeminate." But things in Baghdad were different. These boys didn't have any family to speak of or material possessions—and certainly not a box of toys. When we showed up, the boys grabbed the dolls like they were Tonka trucks. They were laughing and playing and genuinely happy.

Of course, I saved the best for last and pulled out the football. As soon as they saw it, the children who had played catch with me before shouted, "Football!" They remembered! We played catch for a while, and I think somebody also brought out a basketball. But there was something about that pigskin. Funny, with their Islamic beliefs against pigs, I wonder if they would have reacted with such joy if I'd said, "Pigskin"?

Playtime was over too soon, and once again, it was time to go back to the Green Zone. But it wouldn't be too long before I saw the boys again.

In January 2004, Labor Secretary Elaine Chao paid a visit to Baghdad. I had to do all the site surveys and work with her office to plan the trip. When I heard that she planned to visit the al Waziriya Orphanage, I was so excited. I hadn't had a chance to visit the orphanage in a while, so it would be good to see "my boys."

On the day of the visit, we were scheduled to visit both the girls' side of the orphanage and the boys' side. When Secretary Chao walked through the center, the children were on their very best behavior. Sometimes, I wondered what kind of discipline was exercised to keep the kids in line. It being the Middle East, I could only hazard a guess.

As Secretary Chao made her way to the boys' side, I grew quite anxious. We had media with us and all these security guys, and it was just a big scrum. I knew I wouldn't get to play catch with the boys, but I hoped to at least hug them hello.

The boys were lined up and standing at attention as the labor secretary walked into the courtyard. I wasn't too far behind her, so I could see some of the boys and recognized them. The craziest thing happened when I entered the area. One of the boys recognized me, broke rank, and yelled, "Football!" I started to tear up. These boys were standing in a reception line to honor a high-level dignitary, but they seemed more excited to see me and remembered our one connection—football. It really touched my heart. I hugged the boys and then followed the secretary on her tour.

That was the last time I saw the boys. Things got so crazy in the war that I just couldn't make it back over there.

I still think of them today. We had heard that the insurgents were recruiting out-of-work Iraqis and others to fire rockets at the Green Zone for a few bucks. The Iraqi economy was still quite weak, and some were getting desperate. Back then, I worried that once the boys were too old for the orphanage, some bad guy would use Iraqi dinar to entice them.

Even though it's been twenty years, and I am thousands of miles away from Iraq, I still wonder about the boys. I am sure most of them are grown men by now. Are they still alive? If they are, what are they doing? Do they remember the football and the fun we had? Do they have bad impressions of Americans? Did some dirty insurgent terrorist convince them to forget about throwing the football and instead teach them to fire a few rockets?

I will never know.

CHAPTER 8

It's Not Smart to Go Hitchhiking in the Middle East, Especially During a War

> *Truth be told, it's not a very smart idea to go hitchhiking. Period. Whether you are in Kansas, Kuwait, or, in this case, Amman, Jordan, it's still a bad idea. While we were out of Iraq and in neutral territory, it was still an unsafe area. We did something very stupid, yet lived to tell.*

In April 2004, Susan, Kristi, Eric, and I decided to take a little R&R (rest and relaxation) trip and catch a flight from Iraq to Amman, Jordan, which is just north of Iraq.

Just like we did at work, the four of us did our parts to get ready for the trip. I was in charge of getting us on a flight. Kristi was in charge of the hotel, and on top of that, she had a "friend of a friend" who would play tour guide as we traveled around the country (for more on this "friend of a friend," see Chapter 12). Eric would help with bags and be our security boy, and Susan was going to help with whatever.

Traveling in a combat area is not like it is in the United States. You can't just go online, book a flight, drive through a little traffic to the airport, and jump on Southwest Airlines. In a war zone, it's a completely different story, with flight times cloaked in secrecy and departures a virtual mystery until you step on the tarmac. Of course, since traveling is dangerous, we were trying to evade attack every step of the way.

You might be wondering whether, with all the stress, drama, and danger involved, it's even worth it to go anywhere outside the comfort of the Green Zone. Why not stay there for nine months? Simple answer—you gotta live! Soldiers went out every day into Red Zone neighborhoods to work with the Iraqis. We (CPA folks) also went out every day to work with the Iraqis; part of life in OIF is getting out and into the culture. Plus, Susan, Kristi, and Eric had all been pretty burned out from the last few months in Iraq. I had just come back from my mother's cancer surgery in the U.S., and I was emotionally numb. We all needed a break and some new scenery.

Yes, we could go online to book a flight, but to keep our movement secret, we had to go through a specific process. We wouldn't know when our flight departed until the day before, and we wouldn't know which plane was ours until we walked onto the flight line. All these procedures were in place to protect us and, hopefully, ensure that travel information wouldn't fall into the wrong hands. After all, we weren't the only ones traveling around, and all of us needed to follow the same procedures to protect everyone: soldier, civilian, and contractor.

Getting from the Green Zone to the airport was always a death-defying trip. In April 2004, BIAP Road, or as the soldiers called it, Route Irish, was the most dangerous road in the world. It's about eight miles from the Green Zone to the airport, and on a few occasions, the harrowing drive had me gripping whatever I could hold onto until my palms nearly blistered.

BIAP Road was so dangerous because it was the main thoroughfare between the Green Zone and the airport. The flat, four-lane highway made slothful trucks and rhino buses an easy target. Convoys carrying all kinds of supplies traveled the route; soldiers and civilians going to and from the Green

Zone were like sitting ducks in a carnival shooting gallery. Virtually everyone who needed to get from the airport to the Green Zone drove down BIAP Road unless they were a dignitary. Those folks were A-listers, and they were lucky enough to get to travel by helicopter and avoid ground attacks. Coalition vehicles, including Humvees and tanks, driving down BIAP Road would perform evasive maneuvers as they neared bridges.

Sometime in the spring of 2004, if not earlier, the military started bringing in huge, sand-colored monster trucks called "Rhinos" to transport people to and from the airport. I considered myself lucky that I never had to ride in one of those ugly things. Suffice it to say that for us to get to the airport, we had to arrange a ride and make sure we had security.

The fun really began once we made it safely to the airport and boarded the plane for the flight to Amman. Flying in and out of Iraq was damn scary. At night, they kept the lights off on the aircraft while we taxied. Once we were airborne, the pilot made rollercoaster maneuvers called corkscrews to avoid being shot down. I hated this more than anything. After takeoff, we pulled so many Gs going up that it felt like the wrestler Andre the Giant had plopped down on my chest.

Later that night, when we finally arrived in Amman, an older Middle Eastern man showed up on the tarmac and asked for our passports. The process of entering Jordan always made me kind of nervous. We had to pay someone to expedite things, and I was never quite sure, once he disappeared around a corner with our passports, whether he was coming back or if we would ever see them, or him, again.

Thankfully, we did. After getting through the passport process, it was time to begin our R&R and head to the hotel. Kristi had booked us into the Four Seasons, one of the most beautiful hotels in the city. After months of trailers, combat showers in tiny stalls, and plastic forks, we were going to live in style at a five-star hotel. Sweet!

The Four Seasons lived up to its reputation. It was stunning! The opulent hotel had a spacious white marbled lobby with exotic flower arrangements, and everything was dust-free and CLEAN! Though we arrived close to midnight, we decided to splurge on hors d'oeuvres in the hotel restaurant. We were like kids in a candy shop, ordering shrimp and salmon bites with hummus and wine. Then we got the bill. It was more than $200! Still, to this day, the four of us laugh when we talk about how great that food was.

We spent the next few days eating, shopping, exploring, and just relaxing. We went to the mall and enjoyed the freedom of shopping inside a building with escalators, shoe stores, and a food court. It was so much more like home than the souk (Iraqi market) near the Green Zone Café and the 18-wheel trailer that made up our post exchange and shoppette.

We made a trip to the Dead Sea and spent the day floating in the exceptionally salty body of water. We also went on a day trip to Petra. If you've seen the movie *Indiana Jones and the Last Crusade*, with Sean Connery, you might recognize Petra as the tiny, historic village built into a cliff where they go to search for the Holy Grail. At the end of the movie, they ride off on their horses with a pinkish-stone city in the background. That's Petra. It's one of the most amazing places I have ever been to, and it was worth the trip.

What a break from hell.

But in the back of my mind, I kept thinking, *How are we going to get back?* I was worried about making the flight and whether or not we could all get on the same plane. I made several calls to the logistics team in Iraq to find out what day and what time we could return. The thought of our return trip kept me edgy during our adventure.

The night before our trip back to Baghdad, I called the travel guys in Baghdad to figure out when we should report to the airport. He said we should be there at six in the morning, a couple of hours before the flight would actually take off. We weren't excited about the early hours, but we had to do what we had to do.

It was Sunday morning in Amman, and just like six o'clock on a Sunday morning in the U.S., it was quiet. Not a lot of people were moving around. We managed to catch a taxi from our luxury digs to the airport, which was a surprise since it was so early in the morning. When we got to the airport, our contact for military flights said the flight wasn't actually going to leave until the afternoon, and we should come back later.

Kristi, Susan, and Eric were not happy with me. But how was I supposed to know that our flight wasn't going to leave when they said it would? Well, we decided to take advantage of the time and go get something to eat. We had heard that the Intercontinental Hotel offered an incredible Sunday buffet, so we decided to get a cab up the hill and try it out. But there was one problem: it was six on a Sunday morning, and there wasn't a single taxi at the airport—nothing. Everyone was like, "What are we going to do now, Traci?"

About that time, I looked over and saw a person getting out of a small car and waving goodbye to the driver. I also noticed that they were speaking English. I was intrigued and stared at the driver and passenger. The bearded driver looked through his windshield at me, but I just stood there. It was so weird, like I was suddenly paralyzed and couldn't move.

Next thing I knew, he slowly drove toward me, stretched his head out the window, and, in perfect English, said, "Do you need help?" I told him we needed a ride to the Intercontinental Hotel. To my surprise, he offered to give all four of us a ride. I was so excited that I lost all common sense.

"Hey, you guys!" I gleefully shouted to my travel buddies. "He speaks good English, and he'll give us a ride."

Susan, Eric, and Kristi walked over and crammed into the little car. My dear friends left me to get into the front seat with our new "friend."

Our driver—I don't remember his name, but I'll call him "Khalid"—told us he was Palestinian but had been living in Jordan for some time. If you've watched the news but do not have extensive knowledge of Middle Eastern history, like me, getting into a car with a Palestinian in a time of ongoing animosity between Jews and Palestinians would make you a little nervous. On top of that, I noticed that we were driving through a pretty rough neighborhood.

"Khalid" was talking, but I didn't hear a word. My mind was racing. I looked over and saw warehouses and poverty, and I kept thinking, *This is a great place to kidnap a few Americans and dump them!* Then, I forced myself to focus on what "Khalid" was saying so I wouldn't look so nervous.

As I concentrated on "Khalid's" words, I heard him say that Jordan was about 60 percent Palestinian. I didn't know that. Then, sensing my anxiety, he started to talk about the neighborhood we were driving through. He said the city of Amman was made up of eight rings, or circles, with the least wealthy at the bottom, or first ring, and getting wealthier as you went up the rings to the eighth. Something else I didn't know.

By then, I was feeling more comfortable. "Khalid" asked if we spoke Arabic and then mentioned a few useful phrases in the language, such as *"la,"* which means no. He also explained the best way to use it. For instance, if we went rug shopping and wanted to bargain and didn't like the price, we should say, *"La, la,"* wave our hands and walk away.

We then started talking about how well he spoke English. He said he had learned to speak English in school, and his children were learning English in their school as well. He said it was mandatory.

With each ring the car climbed, I began to grow more at ease with our new friend and the cultural exchange.

A short time later, we arrived at the Intercontinental Hotel. "Khalid" was very gracious when he dropped us off. I handed him some Jordanian dollars as a small gesture of our appreciation for the ride and the lesson. But I didn't have enough "JDs" on me to thank him for the priceless life lesson you can learn from anyone: while we lived in different countries, thousands of miles apart, "Khalid" wanted the same thing for his children as any American parent—a good education and a safe environment in which to raise his family. I didn't know "Khalid's" background other than what he shared with us in the car ride. But he was open enough to share with us a few things that opened my eyes to his world.

As soon as "Khalid" drove off, Kristi and Susan almost screamed in unison, "He speaks good English? What the hell was that?"

I reminded them that I wasn't the only one who'd gotten in the car; they'd gotten in with me. We all admitted we had been pretty scared, especially when we'd cruised through the dicey part of town. But all of us agreed we had learned a lot that morning. First, it's hard to catch a cab at the airport in Amman on a Sunday morning. And second, NEVER accept a ride from a stranger near a war zone—even if he speaks good English!

CHAPTER 9

Honey, Er, Sir, I Think I Blew Up Our Car

> *When you're in a combat zone, there aren't many opportunities to laugh. Most of your days are spent dealing with loss or bracing for the next one. But I learned the lesson that no matter how bad the situation is, at some point you gotta laugh. It's like the scene in the movie Sex in the City after Big abandons Carrie at the altar. She's completely shattered, and she asks the girls if she'll ever laugh again. Miranda assures her she will laugh again when something is really funny. Similarly, in a war, a moment will come when you just have to laugh.*

In the war zone, we treated vehicles like some people treat rental cars—we trashed them. We weren't supposed to, but sometimes that's just the way things happened. The military expects that some vehicles are going to be damaged; after all, in a war zone, a rocket soaring through the parking lot could break a window or two.

We considered it a privilege to have the keys to a Suburban in the Green Zone. In the Strat Comm office, we had to share the SUVs since we only had a couple of them. They were used to escort the media or run errands in the area. We were instructed not to drive the vehicles into the Red Zone unless we had weapons or a security team. Since our vehicles were not "hard cars," or bulletproof, they couldn't withstand an attack. Usually, we signed the car out, got the keys from the office manager, and let someone know where we were going.

One night, we were invited out to dinner with some of the journalist folks from the networks. The journalists stayed at different locations in the Red Zone, but most stayed at either the Sheraton or Palestine Hotels just across the Tigris River from the Green Zone. We didn't have a hard car or security team to escort us to the hotel, so we decided to get a vehicle, drive to the checkpoint, park it, and meet the network's security team so they could drive us to the hotel.

Kristi and I and a couple of others piled into the vehicle and drove over to the checkpoint. When we pulled up, there really wasn't any place to leave the SUV. Checkpoint 2 jutted out into the main streets in Baghdad and was always a ripe target for an attack. It had two entrances. One was on the backside of the Iraqi Governing Council's main office. The other entrance was about a block away toward the west and close to the convention center and Rashid Hotel.

The checkpoints weren't exactly built to include stadium parking. Car bombs, or VBIEDs (vehicle-borne improvised explosive devices), were the weapons of choice for insurgents. All an insurgent had to do was fill a car with explosives, drive up to a checkpoint, and blow it up. Such bombs were extremely lethal. With that in mind, the Green Zone security guys didn't like to have too many cars sitting near the checkpoints.

As we pulled up near the checkpoint, I saw a soldier in a tank and asked him if I could park the Suburban somewhere that wouldn't be in the way. He pointed to a dirt lot and said I could probably leave it there. A couple of other cars were already there, and it looked safe enough, so I thought it was okay. Kristi and the rest of us headed to the checkpoint to meet our escort.

We had a fun time at dinner. It was great to get out and forget for just a few moments the dangers of living in a war zone. After a few hours, we gathered ourselves together for the ride back home to the Green Zone.

When our escorts dropped us off back at Checkpoint 2, we got out of the vehicles and immediately detected that something wasn't right.

As we walked toward the checkpoint, we noticed there were a few more tanks around, and a bunch of soldiers were crouched down around them, using the armored vehicles like blast walls. It was really disconcerting.

Suddenly, one of the soldiers yelled at us to get down. We crept over to a nearby soldier and asked him what was going on. He said there was a suspicious vehicle in the dirt lot near the checkpoint, and they were going to blow it up. The next moment, there was a flash of light and a deafening explosion from the dirt lot—right where I had parked the Suburban!

We were like, "Oh shit, they just blew up our car!" We stared at each other, incredulous. How were we… no, how was *I* going to explain to our boss that we had lost a vehicle because we'd decided to go out and leave it behind, only to have the EOD team think it was a car bomb and blow it up? Then, a nervous, uncontrollable laughter hit us.

We weren't just giggling but doubled over laughing. This was just too wild to be true! We were hysterical. If you're an outsider reading this or a critic ready to slam our immaturity, hold on for just a moment. In a war zone, there truly isn't much to laugh about. In this crazy environment, we weren't just caught up in the moment; it was a physical and emotional release. We cried when we were hurt, and that happened almost daily. Rarely did an event occur that was so off the wall that it allowed us to relieve ourselves with tears of laughter. So, call it what you will. Judge us if it makes you feel better. For us, it was an event that brought the IraqPak closer together.

Amidst our side-splitting emotional release, someone had the presence of mind to ask how we were going to get back to our trailers. It was after ten at night, and even though there was a shuttle bus in the Green Zone, we weren't sure if it was still running. Once the laughing stopped, we pulled ourselves together and walked over to the other checkpoint near the Rashid Hotel, where we waited for the shuttle bus.

All the time, I kept saying, "I can't believe I blew up the car." Since the vehicle was in my name, I was responsible for reporting the incident to our supervisor. First, I wanted to find out exactly what had happened. After a long wait, thank the Lord, the shuttle bus came by the hotel to pick us up. We still kept giggling about the absurdity of trying to tell our boss the reason we were one vehicle down was because it had been blown up in the parking lot.

When we got back to the palace, I went straight to the Green Room to 'fess up to the crime. Colonel William Darley, the public affairs officer for General Sanchez, was in the office, where he always was, even though it was well past eleven o'clock. Colonel Darley was a quiet, cerebral officer, and he looked like a soldier who had survived a few battles. Initially, I was afraid of him because he never smiled and always seemed cross. But then, one day, I got up enough nerve to ask him, "Colonel Darley, you're always in the office. Do you ever go to sleep?"

He said very curtly, "Yes… during meetings." I fell in love with him right then, and he became one of my favorite people.

I was relieved to see Colonel Darley in the Green Room. I knew that if I told him what had happened, the reprimand wouldn't be as harsh as it might be with someone else. When I told him the story, he looked over his glasses, smiled, and said, "Okay, let's go to the JOC (Joint Operations Center) and see if they can tell us what happened." The JOC was a headquarters within a headquarters. You needed a special clearance to get access, and I didn't have it at the time. The JOC was much larger than I thought and, with all its bells and whistles, quite intimidating.

Colonel Darley went over to the soldier on duty and very seriously asked if they had any info on a car that had been detonated near the checkpoint.

"Because she"—he pointed at me—"thinks her vehicle was accidentally blown up."

I was completely taken aback when the soldier burst out laughing! Then he turned around to his colleague and said, "Hey, you gotta hear this one!" Then he told me to tell the other soldier what had happened. I felt like a two-headed bearded lady on display at the circus. When I recounted the series of events to the second soldier, they both laughed out loud like they had just heard a joke for the first time. I guess they needed an emotional release, too.

Once the guffaws stopped, they gave me a description of the vehicle that the EOD guys had torched—a yellow car. Thank God! I had been driving a golden Suburban. But then I quickly thought, *What was the damage to the cars that were nearby when the yellow car was detonated?*

The next morning, I got up at about six, even though I had gone to bed late, to walk the mile to the checkpoint and see how bad the damage was. Well, actually, I ran over. As I got closer, I saw the familiar sight of a car that had been blown to pieces. To my relief, the Suburban was still intact. The Man Upstairs had been watching out for me again. There wasn't a scratch anywhere on the Suburban—not one—even though the obliterated yellow car had been sitting right next to it. Those EOD guys were really good.

CHAPTER 10

If Your Tour Guide Tells You Jesus Blessed the Water, Remember... That Was Two Thousand Years Ago

> *Okay, this one is pretty self-explanatory. Remember, when you are in a foreign country, don't drink the water... no matter what the local guy says!*

Anyone who knows me knows that I am famous for saying, "God watches out for fools, children, and Traci Scott." I've also learned in my short time on earth that the Lord has a funny way of doing things. Here's one of those funny times, one that shows how you better be careful what you ask for because you just might get it.

I've struggled with my weight for a good chunk of my life. I was always a healthy girl with an athletic build and a little bit of extra plump on top. I've been anywhere from a size 12 to about a size 6 on a good day, which isn't too bad. But when I arrived in Iraq, everyone was telling me that I would soon be on the "Baghdad diet." The saying is that when you are in Iraq, you either gain three hundred pounds or you can bench it. Lucky me, I was in the Green Zone for just a couple of days before I got slammed with Saddam's Revenge, kind of like Montezuma's Revenge, Iraqi style. It's gastrointestinal distress that keeps you packing Pepto Bismol like you're chugging from a water bottle. Sadly, this was only the beginning.

I am also a sucker for religious artifacts. On this little adventure, my naiveté cost me a great deal more than the Jordanian dollars I shelled out to appease my appetite for theological knickknacks.

Susan, Kristi, Eric, and I decided to take an R&R break and go to Amman, Jordan. While there, we thought it would be kind of cool to take a day trip to Petra. Petra is an ancient city that's built into the side of a mountain. "Petra" means *rock* in Arabic, and the ancient city is truly an incredible sight. As a matter of fact, the blush-colored rock dwelling is one of the New Seven Wonders of the World (according to the New Seven Wonders Foundation).

Kristi had a "friend of a friend" tour guide agree to take us on a day trip. I think he said his name was "Happy," which should've caused some concern, but the guy was always smiling and seemed eager to show us the sights.

We left Amman late in the morning. As we drove out of town and into the countryside, I was amazed at how much Jordan, while thousands of miles away from the U.S., still looked like any countryside in America. We could have been driving through Virginia or North Carolina. That is, until we came upon our first stop. It was the Islamic version of a Stuckey's—a gas station with an Arabic souvenir shop. But this was even better; there were Arabic rugs and all kinds of trinkets. In the back of my mind, I wondered, *How much of a cut did Happy get for bringing in a group of callow tourists?* Nevertheless, I bought a nice little wall hanging, and we were invited to have chai tea. There's another cultural lesson: whenever you are a guest at someone's house, office, or, in this case, Middle Eastern souvenir tourist trap, you are invited to have a cup of tea.

We made our way further up the highway, and then Happy pulled the car over to the side of the road. We all froze. He pointed to an area off the road and kept saying something about the "highway." We didn't budge. We all looked at each other and thought the same thing: *Is he taking us out here to dump us and leave?*

Happy got out of the car and gestured animatedly to the valley. He kept saying, "Come see the highway." It took several minutes before we dared to open the car door. When we finally got out of the vehicle, Happy led us to a ridge overlooking a canyon. He kept calling it the "King's Highway." History books tell us that the King's Highway was an ancient trade route that passed through Petra and went east, apparently all the way to China.

We walked over, did the touristy thing, and took a few pictures. Then we quickly scampered back to the safety of Happy's car with a little more trust in him and a readiness to see what lay ahead on our adventure.

Finally, we arrived in the small town of Petra, which is still a good distance from the rock dwellings, and oh, yeah, we still had to walk an hour to the ancient city. Happy pulled his car over to the side of the street and said, "You must come see Jesus. Jesus was here. Jesus blessed the water."

Now, this piqued my interest. He kept saying, "Jesus was here," as he led us to a tiny concrete building. It didn't have a door, just an open doorway leading into a stone building. It was the craziest thing. A little creek ran through the middle of the room—a waterway in the middle of the building, kind of like a Japanese bathhouse. Happy kept saying, "Jesus. Jesus blessed the water." It was almost like Jesus was standing right there.

Happy ran to the back of the room and pointed at a huge rock. He said the rock was like Jesus. And if you turned your head to the side and squinted your eyes a little, it did resemble the profile of a man with a distinct nose pointing up toward the ceiling.

Next thing I knew, Happy reached down into the cemented pond and said, "Jesus. Jesus blessed the water." He cupped his hand, dipped it into the water, scooped some out, and eagerly sipped it. "Good, good," he said.

Well, you know me and my naiveté. Not wanting to offend our host, I followed his lead and kneeled in the same fashion. I should've just stayed where I was, taken my tourist picture, and been done with it, but no, I had to go all the way. Can I say BIG, HUGE mistake?!

I cupped my hand, reached down, and took a sip. It was the freshest, coolest drink of water I've ever had in my life! No chemical or chlorine aftertaste, either. The water was clean, clear, and so crisp! Someone could've bottled that water and made a fortune.

Immediately, Susan and Kristi brought me back to reality. "Jesus blessed the water, Traci?" I told them it tasted good, but little did I know that the one handful of water would just about ruin the rest of my time in Iraq. I wasn't taken down by shrapnel from a mortar attack or bullets from a terrorist's machine gun. Instead, squirmy little vermin in a single sip of water nearly did me in.

We left the Jesus-blessed room and made our way to another tourist trap to buy some trinkets and gifts. Then it was time to finally go down to Petra. We had a great time walking around, looking at the sites, from the cave dwellings to the camels to the place where virgins were thrown off a cliff as a village sacrifice. Visiting Petra was not on my life list of things to do, but I am so glad I took a side trip there and had another adventure. If you can get there, definitely go see it.

After walking around all day, we were starving and ready for something to eat. Happy took us to a hotel with this absolutely delish brunch buffet.

Knowing it was probably going to be a long time before we had a good meal, we ate like it was Thanksgiving and stuffed ourselves silly. Then it was time to head back to Amman. That's when it hit me.

That night, as we were driving back to the hotel in Amman, my stomach started to gurgle. It wasn't from hunger. Lord knows we had a big enough meal at lunch, and I wouldn't need to eat for a week. Before long, that gurgle escalated into pain. I was starting to hurt. Happy pulled over into a gas station to fill the tank, and wonderful host that he was, he came back to the car with ice cream bars for all of us. But not for me—by this time, I was curled up into the fetal position and leaning against the car door for support. I was feeling it. I didn't want to eat, smell, or even look at a morsel of food.

The blessed water wreaked havoc on my system. For two months, I couldn't eat a solid meal. I subsisted on crackers and water. Whenever I tried to eat something heavier, I was in the bathroom, flushing it through. My friends Susan and Kristi did their best to humor me during this time. Susan, with her typical dry wit, suggested one day, "We should name your parasite. How about 'Faith'?" That worked for me. Whenever I'd see them, they would say, "Hey, Trace. How is your 'Faith' today?" If my intestines were turned inside out, I'd say, "My 'Faith' is pretty strong today." On other days, it was "a little 'Faith.'"

As you can imagine, two months of a cracker diet took its toll on my body. I was losing weight, my clothes were hanging off me, and I looked like crap. One day, I was really feeling worse than usual. I knew I needed help. This was a war zone, and I couldn't just jaunt down to the Quick Care for some medicine. We had two options for medical attention in the Green Zone. One was the CASH, or Combat Air Support Hospital. But we all agreed you had to be in pretty bad shape to go to the CASH, like missing a limb after a mortar attack. I was hurtin', but I could still walk.

So, I finally got myself together and headed to the second option, a clinic in the palace. I told the Army nurse that I had had the runs for about two months and felt awful. He told me that if I didn't start to eat and keep things in my system, they were going to have to send me home. Home? I couldn't go home! It was the end of May, and we were so close to the handover to the Iraqis. I couldn't quit now. That was out of the question! Nope, I was not going to go home! I wanted to stay in the war zone! I made up my mind that I wasn't going to leave.

The nurse handed me a packet of Cipro in a brown envelope. You may remember the name Cipro. A couple of weeks after the attacks on 9/11, some numbnuts decided to fill a bunch of envelopes with anthrax and mail them to a couple of Senate offices on Capitol Hill. The anthrax was so lethal that as the letters made their way through the mail system, it leaked out, killing five people and sickening more than a dozen others. I happened to be working on the Hill at the time, and it was just insane. The anthrax scare forced medical staff to hand out Cipro to congressional staffers who were possibly exposed to the deadly powder. Cipro kills everything.

The Cipro the nurses gave me only took a day or two to kick in, and I started feeling better. But as I said, the illness had already taken a toll on my body. I lost about twenty pounds and went down to about a size 2 or 3. To this day, my gastrointestinal system is still jacked up. I have to be careful about what I eat. If I eat the wrong thing, it runs right on through. But the lesson learned, and the side effect earned, is that at 45 years old, I can still fit into a size 4, or even 2, and all it took was some blessed water and a little "Faith."

CHAPTER 11

Sometimes, You Just Gotta Deal with Geraldo Rivera

> *In a war zone, you're going to run into a Geraldo Rivera: a larger-than-life character used to getting his way, rolling over everything in his path, disregarding anything but feeding his ego, and living by the mantra "The rules don't apply to me." In my case, it really was Geraldo Rivera."*

When I was a kid, I had high hopes and celestial dreams of being a TV reporter. My dad, Jesse Scott, would turn on the news, and I would sit there and watch it, captivated by the man on TV. On Sundays, Daddy used to watch *Meet the Press* and *Face the Nation*, and I would sit down and watch with him. As I got older, every Sunday, we'd talk on the phone and have our own after-the-show debrief.

Later, I loved to watch Geraldo Rivera on *20/20*. At the time, *20/20* was one of the upcoming news magazine shows where journalists spent more time developing and telling a story. Unlike the two-to-three-minute pieces of the evening news, magazine news stories could be 10 to 15 minutes long.

Geraldo Rivera was a new kind of reporter. To me, even in my youth, he was hot! He was this sexy Latino guy with a really thick mustache and a certain suave way about him. At that time, you just didn't see many like him on TV. Most anchors I watched while I was growing up in Kansas City had names like Wendell Anschutz, Larry Moore, or Bruce Rice and had the Midwestern "Dudley Do-Right" look about them. And then there was Geraldo. He wore his hair a little longer, wore leather jackets, and was just plain hot. He eventually left *20/20* under controversial circumstances—okay, I think he was fired. But he went on to do other stuff, and I kind of lost track of him after the Al Capone vault thing.

Years later, when I finally reached my lifelong dream of being a TV reporter, I had a chance to meet him. I was working in Las Vegas as a general assignment reporter for KLAS-TV, and I usually covered the stories nobody else wanted to do (which is also another story). I was covering "Fight Night," a boxing match between two Latino brothers, and Geraldo came over for an interview after the fight. There's always the possibility that when you meet folks you've admired, they may act unkind and dismissive. This time, though, Geraldo was kind enough to do a live interview with me and take a picture afterwards. I thought it was way over the top for him to be so gracious.

By the time I had to work with Geraldo Rivera in Iraq, he had already been kicked out of the country once. At the beginning of the war, in late March of 2003, he was embedded with a unit and drew a map in the sand showing exactly where the unit was located and where they were about to attack—on live TV! This is a huge no-no in war. You don't show the enemy where you are and how you are going to attack them. It's just common sense. Geraldo's credentials were snatched, and he was kicked out of Iraq.

In the spring of 2004, Dan Senor, the spokesman for the Coalition Provisional Authority (CPA), came to me and said they were going to give Geraldo another chance and let him come back to Iraq. My mission was to escort him and assist with stories on the rebuilding of the country. By this time, I was starting to get hardened by the war and had no patience for stupid people or stupid stuff. I'd also found my voice and, more than that, the courage to stand my ground. I was ready for Geraldo.

On the day of his arrival, I received word that he would arrive by chopper, so I drove over to the LZ to meet my charges. Geraldo was going to be a handful. When you deal with someone with his national reputation—and even more inflated ego—you have to handle them a certain way, or the situation can get out of control. Geraldo, suffice it to say, had been around the block a few times—and he knew how to manipulate a person and a situation. I didn't have near his experience.

I kept thinking, *How am I going to get on top of his ego? How am I going to make sure he knows I mean business?* When dealing with personalities like Geraldo's, you can play the role of a meek errand girl, there to serve his whim, or you can be bold and in control. You just have to know who and what you are dealing with and then step into the role.

Geraldo and his brother, Craig Rivera, climbed into the van, along with their camera and soundman and a load of equipment. Craig was Geraldo's producer but also a reporter in his own right. At that moment, the idea hit me: *Be tough. We're in a combat zone.* I had the authority, and I had to assert it right off the bat, or I was going to lose it and them.

They were all settled in the backseat of the van, with Geraldo sitting in the middle. I turned around and said to him, "Hi, my name is Traci Scott. A lot of people have worked really hard to get you here, and if you fuck this up, we will never do anything with you ever again!"

From that day on, I never had any problems working with Geraldo Rivera.

CHAPTER 12

If the Soldier Tells You There's No Room on the Bus, Don't Sneak Through the Back Door

> *Life is a series of near misses. I no longer get upset if I miss a flight, a bus, or a turn on the highway. The lesson here: Don't force your way or throw a tantrum if you don't get the job you want, or the taxi you thought you hailed. Think of the story of the person who missed a train and was late to work at the World Trade Center on September 11, 2001. If you sincerely make an effort to be someplace and think you've missed the boat, it really may be a blessing in disguise.*

Things were rough in Iraq in April 2004. Four Blackwater security guys had been ambushed and cut up, and their bodies had been dragged through the streets of Fallujah. It seemed like there were more attacks than ever on the Green Zone. I had just returned to Iraq after visiting my mom at Walter Reed, where she had undergone surgery for ovarian cancer. Yep, it was a pretty tough time.

The trip Kristi, Susan, Eric, and I took around that time to Amman, Jordan (described in Chapter 8), was one of the best vacations I ever had. We toured ancient biblical sites, ate great food, and somehow managed to get a little rest. By the time the week was over, we were ready to head back home—to Iraq. However, getting home was another story. The day we left Amman, hostilities had escalated so dramatically in Baghdad that I wished we had never left. For the second time in my life, I found myself making peace with death.

We were at the airport in Amman, waiting for our flight, when we received word that it would be delayed due to fighting in Baghdad. Once mortars started flying, there was no predicting how long it would last. Kristi, Susan, Eric, and I were all sitting there, cooling our heels, anxiously awaiting word on the flight and wondering how long it would be before we could board the C-130.

A large contingent of Iraqis was waiting for the same flight. They had been part of an educational program in Amman and were hitching a ride back to Iraq like we were. It was not unusual for Iraqis to travel to Jordan for training and other education seminars. It was the closest and safest location, and it also gave them a break from the war in their front yard. But as we would later learn, they were traveling under a specific coalition-sponsored program and were taken care of while we were left to fend for ourselves.

Finally, after what felt like hours, we were cleared to board our plane. I remember the day being warm, windy, and kind of dusty. We crammed on the C-130, and each of us tried to grab a jump seat amidst the throng. The plane was sweltering, and it reeked of body odor. It was nauseating, but if we wanted to get back to Baghdad, we were going to have to suffer in silence.

During the uneventful flight, I nodded off for a bit, and when I awoke, I realized that something didn't seem right. At that point, there was word that we would have to fly around for a bit. We knew then that it was because BIAP was still under attack, and it was too dangerous for us to land.

When you are on a commercial flight and the pilot comes over the loudspeaker to announce there's bad weather on the ground and you have to fly around until it clears up, you may look out the window or get back to reading your book and not think too much about it. Also, if the weather or circumstances are bad enough, the pilot can talk with air traffic control and figure out another airport at which to land.

We really didn't have that option. There was only one runway close enough to handle a C-130 landing, and that was at Bagram Air Base. It was close enough, but even if we landed there, what was the military going to do with a plane full of Iraqis and other civilians who needed to get to Baghdad? It's not like there was a customer service center on the ground that would get everyone a hotel room for the night. Also, we weren't sure who was on that plane, and it might have been a security risk to have dozens of Iraqis staying unescorted on a U.S. military base.

Suddenly, we felt the plane descend. I forgot to mention that when you are on a C-130, you are more like cargo traveling on a workhorse plane. There aren't accommodations. Seats run along the wall of the plane and down the center aisle of the cargo hold. There aren't rows of seats stretching across the plane, allowing you to recline for a more comfortable flight. You're sitting knee to knee, shoulder to shoulder, strapped in by something that resembles a car seat belt on a seat that's built like an old-fashioned military stretcher. It's not comfortable. On some planes, if you've got to use the "facilities," there is no need for any warnings about getting too close to the cockpit. You walk into a corner, pull a curtain around you, and squat on the folding chair.

Oh, I almost forgot one other thing: there are only about four windows in the C-130, two on each side toward the front and two on each side toward the back. Suffice it to say, you really can't see outside, but on this flight, that was a good thing.

As we were gliding down, the plane started to do some crazy maneuvering, far beyond any rollercoaster ride. If you've landed in Iraq a few times, you can tell by the sound of the engines cutting back and the forward slope of the aircraft when you are about to touch down. Your body leans sideways against the shoulder of your seatmate, and you fight against gravity not to crush them, but the force is too strong. We kept going down, and just when I thought we were about to touch down, the plane abruptly shot back up in the air.

Kristi said, "I just saw smoke!" That sent me into panic mode. It was obvious that Baghdad Airport was still under attack. Our pilot was trying to slide through the attack like a seamstress threading a needle, looking for that one tiny spot of clear air space to squeeze that giant bird through. We were going up, down, and turning, and it was making my stomach churn. I'd flown quite a bit, and this was the first time I seriously started looking down for the airsickness bag. Our Iraqi brethren weren't doing much better. We were all looking around at each other, scared to death.

The pilots banked the plane, turning it around for another attempt to land. The plane thrust forward, and we descended again. I was nauseous. Bracing myself, I thought, *This might be it; I might not make it.* I made peace with things for the second time in my life (the first was while riding on the back of Robbie Knievel's motorcycle, which is another story). Within moments, we hit the ground! We finally landed—safely! Thank God! Hamdallah! Everyone was cheering and applauding. Those pilots were awesome!

When we finally stepped off the aircraft, we saw plumes of smoke surrounding the flight line. While there weren't any fires on the tarmac itself, there were pockets where it was clear something had hit and left behind small puffs of black cloud.

After surviving that nightmare, we now had to figure out how to get back to the Green Zone. Usually, buses transported passengers back and forth, but because of the attacks, the airport had been shut down, and the buses had stopped running.

Kristi started working the phone to see if the Strat Comm security team could pick us up, but our security team was prohibited from leaving the Green Zone. We were stuck.

Kristi, Eric, Susan, and I conferred amongst ourselves and decided we didn't want anyone from our security team to get killed just to come out and pick us up. It wasn't worth it. So, we hunkered down and tried to figure out what to do for the night.

About the time Kristi was calling for a ride back to our version of civilization, we noticed the Iraqis we had been traveling with were heading toward a big, beautiful bus, guided by U.S. soldiers. Susan and I both had the same idea: *Why don't we try to get on the bus with the Iraqis? We are Americans. Surely, the soldiers won't leave us behind if they are taking care of Iraqis!*

We went over and talked to a soldier who looked like he was in charge of the bus. He assured us there might be some room, and Susan and I got excited. But just as quickly as we thought we'd found our way home, our hopes were dashed when somebody said, "There's no room on the bus."

Damn! Susan and I stood there, perplexed but undaunted. We were gonna get on that bus! I have to take a moment to explain my dear, sweet friend Susan. She, like me, is a military brat. Her family is from Nebraska, and she is about as white bread as you can get. Susan personifies Ms. "Goody Two Shoes." She is a tall, striking brunette with a very laid-back, easy going personality. She's highly intelligent and scholarly, but she isn't what one would call "streetwise."

Well, as she and I were trying to figure out how we could sneak onto the back of the bus, a soldier slowly sauntered over to us. In a very low voice, just loud enough for us to hear, he said, "You wanna ride on the bus?" It was what we now call a prison yard moment. A guy walks by and asks if you are interested in the goods. You surreptitiously slip him the cash, and then he nonchalantly walks out of the picture. Nothing more is said.

Well, Susan, being her un-streetwise self, shouted so everyone could hear, "Yes, we want a ride on the bus!"

I almost jumped out of my skin. "Ssshhh. You're not supposed to say anything," I hissed to her.

Well, that didn't work. Let's just say our plot was exposed, and the bus police told us that we would not be riding with the Iraqis back to the Green Zone. We would be stuck at the BIAP terminal for the night.

BIAP's military terminal consisted of several Quonset huts, blast walls, a bank of porta-potties, and the back of an 18-wheeler truck as the PX (post exchange). After all the stress of the flight and trying to find a way home, we realized we hadn't eaten since that morning, so we headed over to the PX to get something to eat.

A few moments later, we had chips, peanut butter, jelly, and crackers and were looking for a place to sit down. The only spot available was next to a set of cement blast walls and the latrines, so we sat down next to the porta-potties and shared our dinner of chips, dip, and crackers. We laughed at ourselves. We'd come a long way from the shrimp hors d'oeuvres and wine we'd enjoyed at the Four Seasons earlier that week.

Later that night, we were finally rescued.

By nightfall, word had gotten around the Green Zone that Kristi, Eric, Susan, and I were stuck at BIAP. Kristi's colleagues at the Ministry of Health called to tell her that their security team was on the way. Now, as you may or may not remember, the "C" in CPA stands for Coalition. Other nations were involved in the rebuilding of Iraq. That was true for the security teams as well. It so happened that the security team for the Ministry of Health was made up of a group of South African mercenaries.

Now, in some circles, "mercenary" is a dirty word. It connotes a bunch of rifle-wielding thugs running around shooting anything that moves. This is not what I experienced with the South Africans I met. If anything, I had my own prejudices toward the South Africans because I had never met any. All I could think of was apartheid and how they would look at me with disdain because I was an African American.

It must have been about ten at night before our heroes arrived in their shining white SUVs. They jumped out of their vehicles and came over to greet us. I had never been so happy to see a group of men in my life!

One of the security guys was a tall, beautifully dark-skinned South African. He walked over to me, put out his hand, and said, "Hello, my sister!" I can't describe the wave of emotion that overcame me. Instead of shaking his hand, I gave him a big hug. The emotion I felt wasn't just because these men had risked their lives to rescue us. I felt as though I had come face to face with my birthright and history. My family traces their roots back to Guinea in Africa, and I had met a few people from Africa at Howard University in D.C. But when he said, "Hello, my sister," it was as if he had reached across the oceans and generations to welcome me to this world.

The rest of the security team greeted us as well. François was the leader of the team. He was, like my blue-black-skinned brother—rather hot, blond, blue-eyed South African Adonis: tall, slender, and packing muscles. I'd seen François around the palace but kept my distance because he looked so intimidating. But at that point, it didn't matter. François and his team had the balls to come get us and take us home, and I admit I wanted somebody with that kind of chutzpah to save us.

Kristi and Susan piled in one truck, and Eric and I hauled our stuff to the other Suburban. We were going to go on one of the most dangerous rides of our lives. The security guys told us things had been bad on the way out, and they would have to do some defensive driving to get us back in one piece. Eric is about six foot two and a good-looking, muscular guy. He's really sweet and kind, too. He and I had worked together but weren't very close—until that ride back to the Green Zone.

When we crawled into the back seat, Eric and I knew this could be the last ride we'd ever make if our team didn't dodge the mortars or gunfire that we anticipated. I put my hand down on the seat of the car, and the next thing I knew, I had reached over and put my fingers on top of Eric's. He reciprocated and grabbed mine. And then we took off.

From our position, it would take about 15 minutes to get down Route Irish and through the Red Zone to safety. After we left the airport and headed down the back roads, the vehicles suddenly took off. Eric and I bounced around in the backseat, holding hands and trying to pretend we weren't scared. We never said a word. I looked over the front seat and saw we were driving about 150 kilometers (90 miles) per hour. We were flying down the highway.

After what seemed like just a few minutes, we arrived at the security checkpoint for the Green Zone—we never heard one shot or saw any mortars. That night, Eric and I bonded in a way that only happens on a battlefield. You survive a crazy moment, and only you and that other person understand it. I developed a new appreciation for Eric and a new love for the South Africans.

When we got back to the Green Room, Susan approached me and asked if I had heard. I didn't know what she was talking about. She said that a bus carrying a group of Iraqis back from BIAP had been attacked—more than 20 people had been injured.

I was so glad we hadn't forced our way onto the bus.

CHAPTER 13

Never Ever, Ever Cry at Work—Ever!

> *My mom told me to never ever cry at work—period! She said that it would make me look weak and too emotional to handle tough situations. However, she did offer me a way out. She said if things got so bad that the tears started to well up, I could go to the bathroom, have my moment, straighten up, and get back into the fight. I must admit, I've done this. But Mom didn't tell me that in a real war zone, all bets are off. One day, overwhelming pain will strike, and there won't be a bathroom anywhere nearby.*

My friend Kristi never cries. She is the toughest person I know. I didn't believe she never cried until I met her parents, and they confirmed that even as a child, she rarely shed a tear. Now, don't get me wrong; Kristi wasn't emotionless, but she would just blow past the rough stuff and push through things.

Kristi and I sat back to back in the Green Room. I was always sliding my chair over to hers to ask her something or say something silly. One day, out of the blue, I slid my chair back to hers and said, "You never cry, do you?"

She said, "No."

And I thought, *Man, she is unbreakable.* This came in handy because Kristi was a magnet for crazy stuff. If something was happening, Kristi was either already in the middle of it or making her way there. But she never broke down.

Once, as she and her crew were driving around Sadr City, a mecca for Coalition attacks, her convoy got lost trying to find the location for an upcoming event. The security folks warned us not to spend too much time circling in certain areas because we'd be a sitting duck for an attack. And, of course, that's exactly what happened. The convoy circled an area too much, giving the bad guys enough time to plant an improvised explosive device (IED). After one too many passes, insurgents detonated the bomb and nearly blew up Kristi's vehicle.

I was sitting in the Green Room when I heard a slight commotion. Kristi came stomping into the room (in her signature high heels) and marched over with this incredulous look on her face.

"We just got IED'd!"

Shocked, I scanned her clothes for patches of blood. She was appalled and angry but not hurt. Other than wondering who on earth would have the nerve to try and bomb her and her convoy, there wasn't a tear, no mascara streamed down her face; in fact, she didn't shed a tear. All she seemed to feel was contempt for the idiots who had tried to take them out.

Kristi worked with the Ministry of Health. She would devise press strategies to help get the word out about what the ministry was doing to rebuild hospitals and clinics, create immunization programs, and address other health-related issues. Her job required her to travel with other staffers outside of the Green Zone to the ministry. At the time, we were under strict orders not to travel into the Red Zone without security. If you worked for a ministry, you got the real security—the tough-guy contractors who were often referred to as mercenaries. The media often demonized the security contractors, especially the Blackwater boys, who did eventually get themselves into hot water.

The security contractors who protected the advisors for the Ministry of Health were from South Africa. Kristi traveled and dealt with them on a fairly regular basis, but I had only seen them in passing and tended to keep my distance. This all changed the night the South Africans came to our rescue.

Kristi had been talking about the South African security team, especially one guy with whom she had developed a friendship, François. She talked about how François did this, and François did that. At one point, I wondered if I was ever going to meet this François. Oh, yeah—she told me he was really cute, too.

When we were stuck at the airport on our way back from R&R, we didn't have any place to stay, and we didn't have any security to get us back to the palace. While we were waiting for the fighting to stop, Kristi managed to get a hold of the Health Ministry advisors. Lo and behold, François volunteered to gather his team and come get us. Finally! Someone was going to rescue us. Also, I was finally going to meet this François Kristi had talked about so much.

My opinion of the South African security team went up a zillion points that day.

When the South Africans arrived, Kristi finally introduced me to François (as I mentioned earlier was quite pleasing to the eye, and Kristi agreed). He looked to be in his late thirties, with blond hair and steely blue eyes, and he was built like a South African cut diamond. If you've seen Daniel Craig, who played the last James Bond, then you've seen François. He was pretty hot. He also had the hardcore personality of Craig's Bond character—cold as ice and very mysterious.

After our brothers from another land safely drove us back to the Green Zone, I was still rather intimidated by François and wasn't quite sure how to thank him for risking his life to come get us when no one else would. I mean, how do you approach someone like that, a trained killer, and thank him for driving like a madman to save us? Saying thanks just wasn't enough. I would find a way to thank him later on when I got the courage to do it.

Several days later, my fear of François completely melted away.

Kristi and other advisors from the Health Ministry were invited to dinner at a local Iraqi woman's house. It turned out not to be a dinner but a feast. Apparently, they had been cooking for days to get ready for our visit. I was always amazed when we went to an Iraqi family's house, and they had prepared a meal fit for a four-star general. Where did they find the food? How did the electricity last long enough for them to cook it? It was amazing, but treating your guests like kings and queens was the Iraqi way.

Another Iraqi culture nugget: their homes are built with the bathroom on the outside of the main house, like an attached garage. At some point, I had to go to the bathroom, so I went outside. I thought I was by myself and had no fear of being alone. When I was finished, I opened the door and almost peed on myself! It was François. Apparently, he took his job quite seriously, as he had stood next to the bathroom door to make sure no one would kidnap me while I was on the commode.

"François! You scared the shit out of me!"

He smirked and then nimbly hopped over a six-foot fence. I could've sworn I heard him laughing on the other side.

Later in the evening, just before we left, I had to make another trip to the bathroom. I knew François would never pull the same stunt twice, so I headed out to the latrine. Once again, I opened the door to walk out, and there was a dark figure standing there with the machine gun! "François, stop it," I snapped.

From that moment on, I could see why Kristi always talked about François the way she did. Not only was he dangerous, but he was very charming.

Whenever I saw François, I would wave and say, "Hey, François!" and he would smile and wave back. But as I said, François was a man of few words. He never conversed, just waved. I would see him every morning in the same place outside of the chow hall, sitting with his team, eating breakfast. It got to the point where I would look forward to seeing the South Africans as I walked out of the DFAC on the way back to the office or wherever I was headed that day.

One morning, as I left the eatery, I looked over at the familiar spot and saw my South African brothers. For the first time, I heard a voice say, "Hello, Traci." It was François! He had actually said hello to me for the first time! How sweet. I was so surprised, and all I could do was wave back and say, "Hello, François." I would've loved to stop and talk, but I had a really busy day ahead of me and didn't have time for chow hall chitchat.

That afternoon, I was escorting a reporter and photographer from *People* magazine. We were on our way back to the Green Zone when I got a call from Kristi. I immediately knew something was wrong because she didn't usually call me; we sat right next to each other and did all our talking in person.

"François is dead."

"No, no!" I screamed. "Not François!"

She told me he had been going to a market to pick up some food for a going-away party for one of the advisors. He wasn't wearing his vest and had been ambushed and shot in the back.

At that moment, all I could think was that I had to get to Kristi right away. Then the tears started to fall. But I didn't care who saw me. François—

who had smiled and said hi to me just a couple of hours before—was dead. *I hate this fucking place!*

But in the midst of my tears, I started to feel another emotion: anger. I was pissed off at François. *Why couldn't he put on his damn vest? Why? Why? If he had just put on his stupid vest, he would be alive.*

When we got back to the palace, I ran into the Green Room and over to Kristi. As soon as I hugged her, she started crying. It wasn't a whimper but a heaving gust of pain. I just held on to her and cried with her.

I learned my lesson that day: on any day—at the most unexpected times—a war zone can break anybody.

CHAPTER 14

It's Not a Good Idea to Wear Flip-Flops to Work: You Never Know When Your Building Will Blow Up in Front of You

> *My mom used to tell me to have some class and dress appropriately, especially when it comes to the right shoes. So here's the lesson: If you are invited to the White House, a combat zone, or a boardroom, dress appropriately!*

What were they thinking?! That was my first thought when I picked up the *Washington Post* and saw the picture in black and white. The Northwestern University girl's lacrosse team was standing for a formal picture with President Bush at the White House, and some of the players were wearing flip-flops! Yes, they wore dresses and dress suits, but the flip-flops just kind of killed the rest of the outfit.

I couldn't believe it! My mother would have marched right into the State Dining Room, yanked the flip-flops off my feet, and burned them right there in the fireplace! I would have been left standing there barefoot. At the time, some of the girls couldn't understand what the uproar was over, since things are a little more casual for this generation—but we all know there are limits.

Note to girls: if you are invited to the White House, combat zone, or boardroom, dress appropriately!

Don't get me wrong. I love my flip-flops, too. I love to wear them anywhere and everywhere (except to work). I love them because they are old, worn in, and comfortable—and they make me feel carefree. I remember when I bought them. I was in Hawaii and needed a pair. I think I bought them in the early 1990s at a drugstore for $6. But as much as I love to lounge around in them, I never wear flip-flops to work. It's just not proper attire for the workplace.

To all the young ladies out there who wonder why I would waste my time writing a chapter on why it's not a good idea to wear flip-flops to work, here's another reason not to do so.

By June 2004, I was so tired of wearing cotton-pickin' Bremer boots. Oh, yeah, Bremer boots are Army boots that the IraqPak affectionately renamed after Ambassador L. Paul Bremer, head of the Coalition Provisional Authority. He wore them 24-7. As a matter of fact, I think he probably slept in them. Anyway, I wore my Bremer boots all the time, and after eight months, I was sick of them.

One Sunday morning, I just had to have a moment of freedom. Jeez, did I pick the wrong day to break my rule and wear my flip-flops to work.

That morning, I decided to go to the chow hall for an early breakfast. I had been escorting a *Time* magazine photographer and needed some "alone time." Escorting a journalist meant spending every waking hour playing "minder" to make sure they didn't sneak around and take pictures of things that would violate op-sec (operations security). Look at it this way—if I had been escorting the *Rolling Stone* reporter who followed General McChrystal, I would've reminded the general that a reporter was there and to keep what he and his colleagues said in check.

I really liked the *Time* photographer, Karen Ballard, but her presence came at a particularly difficult time in my CPA tour. I have to admit I was impatient and sometimes downright surly to her. But the Green Zone and the myriad of unpredictable daily events were starting to wear me down. I was ready to go home, but I still had weeks ahead of me before the handover of sovereignty back to the Iraqis.

Back to my story. It was a deceptively peaceful morning, just the kind of Sunday morning when terrorists most enjoyed firing mortars at the Green Zone. Experience had taught me that if things were too quiet, too peaceful, something was about to happen. However, I had grown complacent as we got closer to the handover and thought I could walk around a combat area without a care in the world.

I got up, put on some clothes, and—feeling rebellious and carefree—slipped on my faded orange, flower-printed flip-flops.

I walked through the sandbags and past palm trees, down the concrete path to the dirt road that led to the palace. *Flip-flop, flip-flop, flip-flop.* I had found the peace I sought with this hundred-yard walk. After glancing at the young Iraqi workers sweeping the persistent dust from the sidewalk, I looked up at the blue sky. *Flip-flop, flip-flop, flip-flop.* BANG! At that moment, my eyes just happened to be fixed in the direction of the red and yellow fireball and the puffy plume of smoke.

I stood dead in my tracks and watched the fire-red and orange plume topped with black smoke spread like a mushroom cloud above the palace. All around me, the workers were jumping into the trenches to escape the dreaded second round.

Then it hit me: *Damn, I'm wearing my flip-flops. I can't go into a building and help anybody out while wearing my flip-flops! What if there's broken glass on the floor? What if the building's on fire? I can't get through all of that wearing my flip-flops! I need my boots!*

I ran back to the trailer. When I opened the door, Karen, the *Time* magazine photographer, asked me what had happened. I was shaking and nervous, and I told her there'd been an explosion and to stay in her room until I came back to get her. I kicked off my ratty flip-flops, found a pair of socks, and shoved my feet into my boots. My hands trembled as I laced them up, tied them, and ran out of the room.

By the time I got to the palace, they weren't letting anyone down the hall near the source of the fireball. I could only get as far as the Green Room. There, I got details from my colleagues. We were all trying to find out if a mortar had hit the side of the building, which would've been the best strike for the nearly incompetent insurgents and their wayward rockets. Some of my Western colleagues used to make light of the poor aim the bad guys had when lobbing mortars our way. But I would always quietly say to myself, *All they need is one good hit.*

That was not the case this time. Actually, it was even more sinister than a rocket attack. According to the rumors, someone had managed to place a bag with an explosive device near the kitchen of the chow hall. I never found out the true cause of the explosion, but that explanation makes sense to me. As I walked in that morning, I never heard the whoosh of the rocket flying over my head. I never saw the sparks or white trail. All I remember are the blast, fireball, and plume of smoke.

Life in the Green Zone had finally hardened me. When I first saw the explosion, I didn't jump into a ditch or dash behind a pile of sandbags. I stood there, trying to figure out if I was equipped to run into the burning building to see if anyone needed help. In a nanosecond, I went through a mental checklist and realized that I didn't have the right shoes to wear into a fire. I was no battle-trained soldier with the daily experience of being under constant attack. I was just a young woman looking for a peaceful Sunday morning walk to work.

After that day, I don't think I ever put on my flip-flops to walk anywhere outside of my trailer.

So, the second lesson here is—don't wear flip-flops to work; you never know when your building may blow up in front of you.

CHAPTER 15

Always Be Prepared in a War Zone Because You Never Know When You're Gonna Have to Put Makeup on the Secretary of Defense

> *The Girl Scout Motto is "Be prepared." Or is that the Boy Scouts? Whatever. When you go into a combat area, you have to be prepared. Have extra water, carry a box of tissues, always have some pain medication. But more than that, learn to be resourceful. If you don't have exactly what's required, look around and use whatever is available to accomplish what you need.*

If you haven't figured it out by now, I am a Black woman. Not that it should matter, but in this lesson from the war zone, it kinda did.

When I was younger, I struggled with my appearance. To put it bluntly, I had terrible acne. It was so bad when I was in high school that one of my classmates called me ugly. Secretly, it absolutely destroyed my self-esteem. But I guess I was resilient. I found other ways to excel, like in sports and school government.

My mom, on the other hand, was tall, beautiful, and always looked her best. She'd even dab on some lipstick if she were going to the 7-Eleven to pick up some milk. Hence, the battle between me and my mom: she wanted me to be a little princess, and I wanted to be a tomboy. Mom was always trying to get me to wear makeup, which I hated. When she won, and I had to wear it, I always felt like it made things worse.

But then I discovered that I wanted to be a TV reporter. There's no way you can make it on TV without some heavy pancake to cover the flaws. I finally raised the white flag, made peace with the idea of makeup, and started to wear it. I decided that if I was going to wear the stuff, I needed to learn how to do it the right way. When I worked at *CBS This Morning*, I used to watch the makeup lady, Christie Brown, apply makeup to guests on the show, many of whom were Washington and Hollywood celebrities. Sometimes, when she was done, I'd ask her about why she used one color over another, why this method and not another one, and other tips.

When I finally got my first job on TV in Las Vegas, I decided to take a trip to New York and stop by the MAC store for a makeup tutorial. I never realized the little lessons I learned would provide me with skills that would come in handy beyond touching up my face.

Fast forward to December 2003, in Baghdad.

A few weeks before, I had spent several days in the Kurdish area in northern Iraq, escorting *NBC News* correspondent Bob Arnot through Mosul and Kirkuk. Dr. Bob, his crew, and I were back in the Green Zone when we learned that Secretary of Defense Donald Rumsfeld had made a quick visit and was going to have a news conference in the palace.

I had heard stories about Secretary Rumsfeld. They weren't all positive. I'd heard he had a temper, was known to be prickly, and quite blunt with people. I'd seen him dress down reporters during Pentagon briefings and was thankful I hadn't ever had to deal with him. Since Dr. Bob was there, I found out where the press conference was going to take place and rushed Arnot and his crew to a back room with the rest of the media. I also heard that someone

I knew would be traveling with the secretary and thought I would just go say hi to my friend and get out of the way.

We were shown to a back room in the palace, where reporters were corralled for the news conference. I escorted Dr. Bob and his crew into the cramped room with the other reporters and helped them set up before Secretary Rumsfeld arrived.

In a side room, I saw my friend, who was a part of the travel party, and I walked over to give him a hug. For some reason, I looked right past the secretary of defense, who was standing a foot away. It was a bit of a chaotic scene. People were rushing around in the room next door to get set up for the media availability, and more people filled the room where they were prepping the secretary. Things were abuzz.

Finally, my friend turned and introduced me to Secretary Rumsfeld. He put his hand out to shake mine. As far as I was concerned, I had met him, and that was that.

Next thing I know, I hear this voice from the heavens above say, "Traci, do you have any makeup for the secretary?" My brain almost exploded! In a nanosecond, several thoughts raced through my head: *Me and my mahogany, NC-50 MAC-wearing Black skin have makeup for the whiter-than-white secretary of defense? In the middle of a war zone? Are you kidding me? What am I supposed to do, slide down to Sephora, pick up some NARS tawny foundation and blot it on him? Have they lost their minds?* My brain was on fire, trying to figure out how to get out of this one.

But then I came to my senses and told everyone to hold on for a second. I remembered that the correspondent, Bob Arnot, should have some. God, I hoped he had some. I darted over to Dr. Bob, who was rushing around in the next room, trying to get ready for the news conference. The words came tumbling out of my mouth, but I finally got it together and asked him if he had any makeup on him. Thank God he did!

As I sprinted back to the other room, I thought, *I saved the day! I managed to find some makeup for the secretary of defense!* As far as I was concerned, I'd done my job, and it was all over with. I ran into the crowded room to hand somebody the makeup.

Next thing I know, that voice from heaven came booming again: "Well, Traci, aren't you going to put it on him?" Now my head was about to spin off my neck! *Are you kidding me? You want ME to put makeup on the secretary of defense? No less than Secretary Rumsfeld?!* In an instant, I thought, *Traci, this is a war zone. You better buck up and do it.* Meanwhile, Secretary Rumsfeld was standing there, waiting.

I thought, *Well, here goes nothing.*

And the conversation went a little like this. "Mr. Secretary, my name is Traci Scott. I am a civilian working here in Iraq, and I need to put a little makeup on you. Trust me, I know what I am doing. I used to be on TV, and I just need to get rid of a little shine you have on your face."

He looked at me, and a smile came across his face. "Just don't get rid of the scar on my forehead," he said. And then I noticed it: a four-inch scar running diagonally across his forehead. I had never seen him in person, let alone gotten close enough to see that big, fat "Harry Potter" scar imprinted across his head!

As I reached up to pat on the makeup, he told me the cutest story about how he got the scar. He said he was out with a girl and trying to impress her, so he dove off a boat dock and slammed right into a rock in the water, splitting his head wide open. As he told me the story, he kinda giggled. I thought, *I can't believe I am standing here putting makeup on the secretary of defense in the middle of a war zone, and he's telling me about his romantic escapades!* It was pretty wild!

He eventually asked me where I was from and why I wanted to come to Iraq. I told him that I didn't know how to shoot a gun or rebuild a country, but I wanted to do my part to serve my country and help where I could. He gave me a very sweet compliment, and I thought, *What a charming man.*

In the end, I think he was okay with the makeup job. He didn't complain about it, and I didn't get kicked out of Iraq for screwing up the Secretary of Defense's makeup… so I lived to see another day in the war zone.

Epilogue

It's been twenty years since I survived Iraq. More than 4,400 military members did not.

While I was happy to return home, I was not the same person. I was angry all the time. I had lost touch with my emotions. I could not cry. I was impatient. I snapped at everyone and everything. Paula, who was still recuperating from multiple surgeries to save her arm, stayed at my townhouse in Maryland. I snapped at her. But she rolled with it as she tip-toed around me.

I was in a deep, dark place, and I didn't know how to get out of it, nor did I tell anyone about it. Although I am sure many knew. I also kept having this agonizing recurring nightmare. I was in a Chinook with others, and it crashed into a field. We crawled out of the wreckage, and insurgents started chasing and shooting at us. I would wake up sweating like I had really been running. I avoided sleep so I wouldn't fall back into the same illusion.

To add to my emotional turmoil, my mom was battling cancer. She died two years later. My stepdad, PopPop, who championed my decision to go to Iraq, joined her in 2021. Now Daddy, Mom, and PopPop are all buried at Arlington Cemetery along with Chad and so many other Iraq War veterans.

The IraqPak all made it home safely, and we have had several reunions to commemorate our role in what Ambassador Bremer called "a noble cause." Many are married, have children, and are successful in their pots-Iraq lives.

Paula Wikle had more than a dozen surgeries to save her arm. But she is doing well and about to retire from the State Department.

Susan, Kristi, and I are closer than ever. While we are all quite busy, we still manage to spend birthdays together and occasionally travel to wherever our whims take us. Yes, we reminisce about our days in "the sandbox," but the funny thing is each time we talk about an event, a new detail is unveiled. To which I always reply, "You buried the lede!" Our friendship was forged in the crucible of war, and it is purer, richer, and deeper than I ever imagined.

Iraq is still standing. But the current political situation is still somewhat fragile, as it is throughout the Middle East. However, I recently saw a story on the news that gave me pause. An Iraqi woman was doing an interview about her post-war life. Behind her was a row of restaurants with lights, music, and a sense of festiveness. She was happy. I sat there for a moment, picked up the remote, and changed the channel.

So, I suppose you may want to know if it was worth it. I can only speak for myself. I hate war. I hate fighting. I hate pain. I hate death. I hate losing so many friends and acquaintances.

And this leads me to my original thought. When I first came home, I went to Arlington to visit Chad. He's in Section 60, which is where Iraq and Afghanistan war veterans have been laid to rest. When I first went there in 2004, his row was the last one on several acres of sloping green grass. Today, hundreds of white headstones dot the landscape - it's full. I have yet to make peace with that.

SNAPSHOTS OF A JOURNEY

This is my mom, Shirley Jones, dropping me off at Dulles Airport on my way to Iraq.

PopPop, Joe Jones, hugging me goodbye at Dulles Airport on my way to Iraq.

Saddam's Palace in the Green Zone: This is before the coalition removed the statues of his head. Coordinating the media to view the removal of the statues was one of my projects as Special Projects Coordinator.

When the day came to begin removing the Saddam statues, many wisecracked, saying, "Heads will roll."

Some early members of the Green Room or Strategic Communications staff. The swords were an iconic location to take pictures when the mortar attacks were in a lull.

One of the most popular ironies on the grounds of Saddam's Palace.

This is Sergeant Stephanie Clark, aka SGT Hulkette; she was one of the first military personnel I met when I arrived in Iraq.

This is one of the buildings in the Green Zone that was obliterated during the U.S. military's Shock and Awe campaign.

The backside of Saddam's Palace. Also the hub for the Coalition Provisional Authority (CPA). Photo courtesy of Brian Leventhal.

Trailer life in the Green Zone. However, this photo was taken towards the end of CPA's work, when things were much more chaotic and dangerous. Hence, the numerous sandbags. Photo courtesy of Brian Leventhal.

The sandbag wall went all the way up to our front doors.
Photo courtesy of Brian Leventhal.

This is what it's like flying behind the gunner, looking out of the doors across Baghdad.

This is Baghdad... Yes, those are satellite dishes on top of roofs.

Some of the members of the IraqPak enjoying a moment together. Front row left to right: Victoria Whitfield (one of the CPA members from the U.K.), Dallas Lawrence, and me. Back row: Susan Phalen and Al El Sadr.

The air traffic control center for the Washington Landing Zone is amazing for what the military accomplishes with few resources.

Kristi and me on the other side of the air traffic control tower at the landing zone (LZ).

My room in the trailer, also known in military parlance as a "hooch."

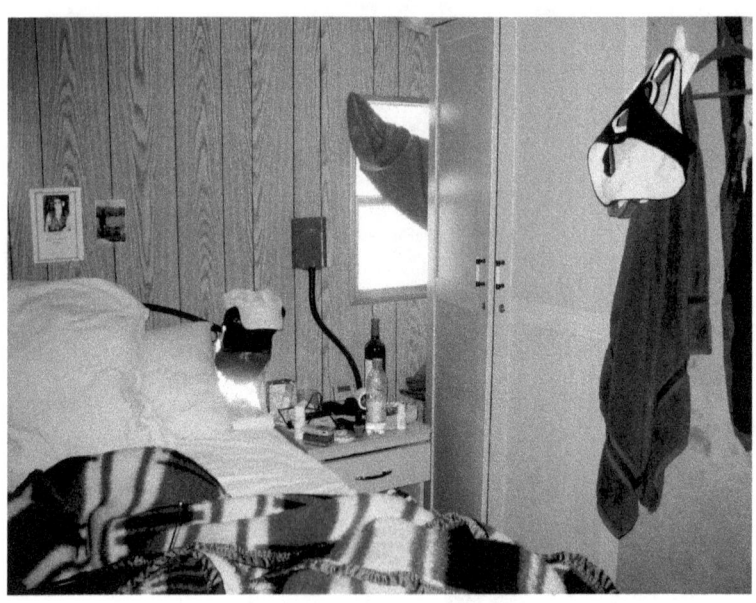

I lived on this side of the "hooch" for nine months.

The trailers had two rooms, one on the right and one on the left, and a shared bathroom split the two sides. The other side, where the TV is, belonged to my roommate. I had five roommates in five months. I needed therapy for that alone.

Iraqis found work helping around the Green Zone. These two became friendly. Every time they came to take out the trash in our trailer, I would always say, "Five minutes! I need just five more minutes before you can take it out." Every time after that, when they would see me, they would shout, "Five minutes!"

This is the famous "Green Zone Cafe." An Iraqi transformed a gas station into one of the few eateries outside of the chow hall we could go to because it was inside the compound. Insurgents blew it up several months after I left.

Shane Wolfe, one of the IraqPak members, and I were so excited to see green grass that we took a picture with it. It's the small things that brought us joy in a war zone.

I served as the Special Projects Coordinator for the CPA. One of my projects was coordinating an interview with Ambassador Bremer for People Magazine. A novelty during our time in Iraq were these 'Bremer Boots.' They were regular Army boots, but somehow they became a fashion statement; thus, Amb. Bremer managed to include them in the interview/photo session.

The Rashid Hotel.

Lieutenant Colonel Chad Buehring. He was posthumously promoted to Colonel o after he was killed in the Rashid bombing on October 26, 2003.

A soldier stands guard over Chad's casket at Arlington Cemetery. Photo courtesy of my mom, Shirley Jones.

My mom and PopPop were sweet enough to attend Chad's funeral while I was still in Iraq. Here they are with Chad's wife, Alicia Buehring. My dad is buried just out of frame to the left. Mom and PopPop are now buried a few sections over from here.

The Army memorialized Chad by naming one of the camps after him.

On October 26, 2008, five years to the day that Chad died, I ran the Marine Corps Marathon in Washington, D.C., in his memory. Anytime someone from the crowd would shout, "Go, Traci, for Chad!" I would tear up.
Photo courtesy of Marathon Foto.

The program from the memorial service for François—the second biggest heartbreak during my time in Iraq. He was killed in an ambush.

One lighter moment: a visit to the Iraqi police academy. I wonder if any of these men are still alive.

One of my favorite moments: hanging out with some young Iraqi orphans. Even though we did not speak the same language, they all understood "selfie."

Susan, with some of her young friends.

*Kristi had something that really grabbed the attention of these children.
Photo courtesy Susan Phalen.*

Another fun time interacting with some Iraqi women was sharing a little of our "Western" culture with them. Here they are, quite intrigued with the game Operation. Photo courtesy of Susan Phalen.

We occasionally had distinguished visitors drop in to boost morale. In this case, it was the former 'Late Night with David Letterman' host himself. Kristi and I decided to get his attention by making up a sign that said, "Dave is our Baghdaddy."

Our job in the Strat Comm office was coordinating media events. On this occasion, we had an event that generated a lot of news for the day—both Iraqi and Western media.

Another chaotic moment with the Iraqi media, this time with former Prime Minister Ayad Allawi. Photo courtesy unavailable.

Another VIP journalist, Peter Jennings, former ABC News anchor. Here he is holding a Flat Stanley for Susan's nephew. Photo courtesy of Susan Phalen.

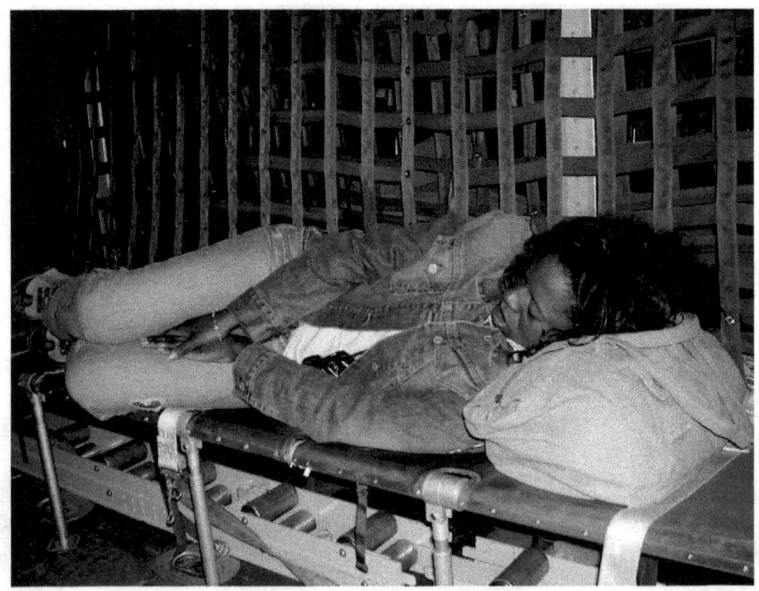

Military aircraft don't have rows of seats like regular planes; they have "jump seats." In this case, I much preferred it.

Brigadier General Mark Kimmitt, spokesman for the military, and Dan Senor, spokesman for the CPA, along with BG Kimmitt's PSD.

In April 2004, Susan, Eric Jewett, Kristi, and I took an R&R break in Amman, Jordan. It turned into quite an adventure. We got on the flight to head back to Baghdad but had a little trouble getting home.

Oh yeah. This is the moment where I got my faith (see Chapter 10).

We were delayed leaving Jordan because BIAP in Baghdad was under attack, as evidenced by the smoke in the background, but we managed to land. However, the airport was still too dangerous for our security team to drive out and pick us up.

This is us trying to make do at BIAP with no way back to the Green Zone. We went from a four-star hotel in Jordan to eating snacks on top of a suitcase next to the porta-potties.

Kristi on the phone trying to secure a ride back to the Green Zone. Eventually, her detail, the South African PSDs, saved us. The ride back was harrowing.

Ambassador Bremer was thoughtful despite the intense pressure. He went to an event and was given a bunch of flowers. He shared them with me and Susan. It was such a fun moment.

I had the opportunity to meet General David Petraeus when he commanded the 101st Airborne while they were in Mosul, before this photo was taken. When this photo was taken, I was invited back to Baghdad to work on his staff for a short time during the "Surge."

Take off Your Hoop Earrings Before Putting on Your Gas Mask

This is me with former Defense Secretary Donald Rumsfeld, years after I left Baghdad. He died about a year after we took this picture.

As we reached the end of our time in Baghdad, Ambassador Bremer hosted a get-together with the IraqPak.

Finally out of Baghdad!! This is after we landed safely in Germany. Our job was done.

THANK YOU FOR READING MY BOOK!

I would like to keep the conversation going!

Scan the QR code to get connected

I appreciate your interest in my book and value your feedback as it helps me improve future versions of this book. I would appreciate it if you could leave your invaluable review on Amazon.com with your feedback. Thank you!

www.ingramcontent.com/pod-product-compliance
Lightning Source LLC
Chambersburg PA
CBHW051836090426
42736CB00011B/1840